SyncML: Synchronizing and Managing Your Mobile Data

Uwe Hansmann
Riku Mettälä
Apratim Purakayastha
Peter Thompson

Foreword by Phillipe Kahn

ISBN 0-13-009369-6

PRENTICE HALL PTR
UPPER SADDLE RIVER, NJ 07458
WWW.PHPTR.COM

Library of Congress Cataloging-in-Publication Data

SyncML: synchronizing and managing your mobile data / Uwe Hansmann ... [et al.].
    p. cm.
    Includes bibliographical references and index.
    ISBN 0-13-009369-6
    1. Synchronous data transmission systems. 2. Wireless communication
systems--Standards. 3. Mobile communication systems--Standards. 4. Computer network
protocols.

TK5105.4 .S96 2002
621.382'12--dc21   2002074853

Editorial/production supervision: *Donna Cullen-Dolce*
Acquisition Editor: *Mary Franz*
Editorial Assistant: *Noreen Regina*
Marketing Manager: *Dan DePasquale*
Manufacturing Manager: *Alexis Heydt-Long*
Cover Design: *Talar Boorujy*
Cover Design Director: *Jerry Votta*
Interior Design: *Gail Cocker-Bogusz*

© 2003 Uwe Hansmann, Riku Mettälä, Apratim Purakayastha, Peter Thompson
Published by Pearson Education, Inc., Publishing as Prentice Hall PTR
Upper Saddle River, NJ 07458

All rights reserved. No part of this book may be
reproduced, in any form or by any means, without
permission in writing from the publisher

The SyncML Initiative has granted to use the SyncML copyrighted material in this book.

Prentice Hall books are widely used by corporations and government agencies for training,
marketing, and resale.

For information regarding corporate and government bulk discounts please contact:
Corporate and Government Sales (800) 382-3419 or corpsales@pearsontechgroup.com
Or write: Prentice Hall PTR, Corporate Sales Dept., One Lake Street, Upper Saddle River, NJ
07458.

Printed in the United States of America

10 9 8 7 6 5 4 3 2 1

ISBN 0-13-009369-6

Pearson Education LTD.
Pearson Education Australia PTY, Limited
Pearson Education Singapore, Pte. Ltd.
Pearson Education North Asia Ltd.
Pearson Education Canada, Ltd.
Pearson Educación de Mexico, S.A. de C.V.
Pearson Education–Japan
Pearson Education Malaysia, Pte. Ltd.

Contents

Foreword   xi
Preface   xiii

INTRODUCTION   **1**

Chapter 1
An Introduction to Data Synchronization   3

The Different Topologies 4
   One-to-One 4
   Many-to-One 5
   Many-to-Many 6
   Hybrids of Many-to-One and Many-to-Many. 7
Summary 8
The Different Usage Modes 9
   Local 9
   Pass-Through 9
   Remote 10
Challenges with Data Synchronization 11
   Conflicting Updates. 11
   Conflict Detection 14
   Conflict Resolution 15
Related Work. 16
   Infrared Mobile Communication 16
   Wireless Application Protocol. 17
   Third Generation Partnership Program 18

Part I SYNCML OVERVIEW   19

Chapter 2
SyncML: An Introduction   21

The SyncML Initiative 24
An Overview of SyncML 26
    SyncML Basics 28
    Additional Key Considerations 30
From an Initiative to a De Facto Standard 33

Chapter 3
SyncML Applications   35

Coordinating a Busy Family 37
    Application Setting 38
    Application Logic 39
    Usage Instances 40
    The Benefits of SyncML 41
Supporting Roving Nightingales 43
    Application Setting 43
    Application Logic 45
    Usage Instances 46
    The Benefits of SyncML 48
The Reach of SyncML Applications 51

Part II SYNCML IN-DEPTH   53

Chapter 4
SyncML Fundamentals   55

The Design Goals of SyncML 55
    Effectiveness over Wireless Networks .57
    Support Transport Heterogeneity 59
    Support a Rich Set of Networked Data 61
    Neutrality to Programming Environments 63
    Support Multiple Synchronization Topologies 64
    Address the Resource Limitations of a Mobile Device 65

Allow Building of Scalable Servers 68
Build a Secure Synchronization Platform 70
Build Upon Existing Web Technologies 71
Build a *Working* Specification 71
Promote Interoperability 72
Architectural Insight into SyncML 72
The Meaning of Synchronizing Two Datastores 74
Language of Synchronization: XML and MIME Usage 75

Chapter 5
Synchronization Protocol   77

Overview 77
Relation to the Representation Protocol and Other DTDs 78
Entities Using the Synchronization Protocol 78
Supported Synchronization Scenarios 80
Phases of Synchronization Protocol 82
Initialization 83
Data Exchange 88
Completion 89
Server Alert 90
Transferring Large Amounts of Data 91
Large Object Delivery 91
Maximum Size of SyncML Messages 92
Multiple Messages per Package 93
Mapping of Identifiers and Slow Synchronization 94
Nature of Identifier Mapping 95
Slow Synchronization 97
An Example Synchronization Dataflow 99

Chapter 6
Representation Protocol   103

Identifiers in SyncML 103
Target and Source Addressing 104
Target Address Filtering 105
Operations in SyncML 106
Modifying Data 106
Adding Data 106
Deleting Data 107
Refreshing Data 107

Searching for Data...108
Grouped Operations on Data 108
Representation Protocol Elements 109
The Message Container Elements 109
The Protocol Management Elements..............................111
The Command Elements......................................112
The Common Use Elements 121
The Data Description Elements.................................131
Text and Binary Representation 133
Static Conformance Requirements 133

Chapter 7
Supportive SyncML Components — 135

SyncML Architecture and Components 135
Complementary DTD Components...................................138
Meta Information DTD......................................138
Device Information DTD.....................................144
Transport Protocols for SyncML....................................148
HTTP Binding 148
WSP Binding..149
OBEX Binding 150

Chapter 8
Security and Authentication — 153

SyncML Authentication..153
SyncML Client/Server Authentication 153
Datastore Authentication 154
Object Authentication.......................................154
SyncML Authentication Types 154
Basic Authentication.......................................154
MD5 Authentication.......................................155
Base64 Encoding...156
Secure Transport..157
Secure Sockets Layer (SSL).......................................158
The SSL Protocol 158
SSL Cipher Suites 159
The SSL Handshake 159
More Information on SSL and Certificates 160

Chapter 9
Device Management   163

Rationale and Overview 164
   Benefits to Interest Groups. 167
   Usage Models. 168
SyncML Device Management Technology 170
   Comparison with SyncML Synchronization Framework. 172
   Bootstrapping in SyncML Device Management 173
   SyncML Device Management Protocol 175
   Device Description Framework. 181
Summary and Next Steps 183

Part III BUILDING SYNCML APPLICATIONS   185

Chapter 10
SyncML API and Reference Implementation   187

Functionality 188
Architecture. 189
Installation 192
Initializing the Reference Implementation 193
Generating a SyncML Document. 194
Parsing a SyncML Document. 198
Communication Toolkit API 201
   Using the Communications Toolkit API 202
The Future. 204

Chapter 11
Mobile Devices and SyncML   205

Wireless and Mobile Characteristics. 206
SyncML Client Architecture and Implementation 207
   Main Technical Requirements 210
   Minimal SyncML Client Implementation. 212
   Mobile Software Platforms 213

SyncML Enabled Applications 214
   Application Types 215
   SyncML Requirements of Applications and Datastores 218
Summary 220

Chapter 12
The SyncML Server   221

A Generic SyncML Server......................................222
   Protocol Management 224
   Sync Management 226
   Data Management 228
   An Illustrative Dataflow 229
Data Paths in Synchronization 230
   Single-Path Synchronization 231
   Multiple-Path Synchronization 232
Functional Expectations from a SyncML Server 236
Performance, Scalability, and Reliability 238
   Exploiting SyncML Characteristics 238
   Exploiting Back-End Characteristics 239
   Exploiting Application Characteristics 239
   Effective Use of Concurrency and Asynchrony 240
   Failure and Recovery 241

Chapter 13
Interoperability Verification   243

Conformance Testing 244
   SyncML Implementation Conformance Statement (SICS) 246
   SyncML Conformance Test Suite (SCTS) 246
Interoperability Testing at SyncFest 252
Virtual SyncFest 254
SyncML Interoperability Reference Pool 254
Recertification 255

Part IV SUMMARY AND THE FUTURE — 257

Chapter 14
Summary and the Future — 259

- SyncML History 259
- Current Market Status 260
- Future SyncML Activities 261
- Future Markets 262

Part V APPENDICES — 263

Appendix A
Bibliography — 265

Appendix B
Glossary — 271

Appendix C
Trademarks — 277

Index — 279

Foreword

With the wireless revolution, we now have the ability to be productive anytime, anywhere. That's because we can get to information anytime, anywhere. Working with out-of-date information is like reading yesterday's papers. We want our information to always be up-to-date, any time, anywhere.

From a technology perspective, delivering this solution to customers presents numerous challenges. For example, we live in a world of heterogeneous devices connected through wireless and wireline networks all supplied by different vendors and/or organizations. There is only one way to make sense of such a complex world: the creation an open standard for synchronization, across networks, across devices. That is exactly what SyncML® does!

Our future is pervasive. To accelerate into this future, the platform provided by SyncML sets up a tremendous opportunity worldwide. That's because it is delivering on the vision of having the right information, anywhere, anytime.

–Philippe Kahn

Preface

Information is the nucleus of today's interconnected economy. We need to be able to access our personal and business information quickly, efficiently, and securely, at any time and at any location, even if this location has no capabilities for wireline or wireless networks to access remote servers.

Pervasive devices like PDAs and mobile phones enable us to remotely use services provided by servers. With these devices we want to access personal or corporate data that we need for our daily business, like contact information of customers, pricelists, or other corporate data.

Data is primarily stored on servers or desktop computers. The mobile device contains a copy of this data and the user needs to be able to manipulate the data on the mobile device. The resulting changes need to be reflected in the source database, without retyping them.

At the same time somebody else could make changes to the data on the server directly. Having two or more copies of a dataset and being able to manipulate the copies independently requires synchronization to reconcile the different changes to form one consistent dataset.

To make all this happen seamlessly, the industry is currently converging upon the required standards. The leading companies in the Information Technology and Mobile Communications industries, like Ericsson®, IBM®, Lotus®, Matsushita®, Motorola®, Nokia®, Palm®, Psion®, and Starfish® founded the SyncML® Initiative as a forum for the definition and promotion of a universal synchronization de facto standard based on XML. Since the first meetings in 2000, SyncML has received a lot of support and the resulting specification has been a great success. All leading companies interested in data synchronization actively participated in the definition and enhancements of the specifications. More importantly, a number of SyncML compliant servers and devices are already available in the market.

About This Book

This book explains the fascinating characteristics of data synchronization and device management and the solution that SyncML offers to achieve global and interoperable synchronization and device management with first-class out-of-the-box experience.

It comprehensively covers the SyncML specifications and the reference implementations. It describes the ideas and design principles behind the specifications and how they evolved to the current state. There are separate chapters dedicated to the integration of SyncML into mobile devices as well as synchronization servers.

This book has a strong emphasis on the challenges that universal synchronization faces as well as the solutions and opportunities that a global synchronization specification offers. One example of these new opportunities is the new generation of services, which deliver added value to a growing number of customers. There are many industries (such as e-business, private or home, finance, or travel) that will use synchronization as the base infrastructure to efficiently exchange data.

This book contains everything you need to know about SyncML and synchronization. The major goal of this book is to make it easy to understand SyncML, its benefits, and to implement it in solutions, on the client as well as on the server side.

The Audience of This Book

Giving a comprehensive and profound overview of synchronization and especially SyncML makes this book very valuable for a wide audience of readers. Following the main thread of this book a reader will find an easy and quick entry to the topics of his interest.

Business managers and managers responsible for IT systems will learn what impact SyncML and synchronization itself has on economy and society. The knowledge about synchronization paradigms, new business models, and a new generation of applications will affect their work as well as their decisions. They will see where synchronization can help businesses to offer new services and new products or how to improve existing businesses to reach a new range of customers.

Software architects and project managers extending their e-business activities to a new front-end using synchronization will read which components SyncML based solutions are made of and how these building blocks are related to each other. This book gives an overview of state-of-the-art synchronization technology and shows which components are available and what needs to be done to build a complete solution.

Application developers getting involved with SyncML will find an in-depth guide through the different SyncML specifications and paradigms, what needs to be implemented, and how to get a perfectly working, efficient, and scalable SyncML synchronization solution. They will get a very good introduction before digging into the details. They will learn how to rapidly enable applications for the use of SyncML and how to avoid usual pitfalls.

No Need to Read the Whole Book

Most of us no longer have the time to read a book cover to cover. Therefore, we have broken this book into parts that may be read in almost any sequence:

- "Introduction"
- "Part I: SyncML Overview"
- "Part II: SyncML In Depth"
- "Part III: Building Synchronization Applications"
- "Part IV: The Future"
- "Part V: Appendices"

Introduction

In this part, we set the stage by providing an overview of data synchronization in general.

Chapter 1: An Introduction to Data Synchronization

This chapter gives a general overview on data synchronization. It introduces the reader to the basic concepts of synchronization and their advantages or difficulties. It describes its challenges and why SyncML is needed. An overview of the other data synchronization standards, like those used in IrMC or Bluetooth™ completes this introductory chapter.

Part I: SyncML Overview

In this part, we give an overview of SyncML, its fundamentals, and typical applications that benefit from using SyncML.

Chapter 2: SyncML: An Introduction

This chapter provides a look behind the scenes of SyncML. It explains what the business motivations were and still are and how it all began. It gives a brief history of the formation of the group, how SyncML evolved from a group of companies jointly developing and promoting SyncML to a real industry initiative with many companies in the IT industry being a member of it.

This chapter introduces the basic concepts behind SyncML and how SyncML fits in a data synchronization solutions.

Chapter 3: SyncML Applications

In this chapter we walk through two applications that SyncML is commonly used for: Personal Information Management, Data Management, and Enterprise. The applications also illustrate the use of device management. It provides insight into how SyncML is used in these applications and how the various players (server and device manufacturers, and most notably, customers) benefit from it.

Part II: SyncML In-Depth

This part gives in-depth insight into the core specifications that the SyncML standard consists of. It provides valuable information how SyncML can be successfully used and gives insight into the reasons behind different design decisions.

It also explains the knowledge that one needs to build an efficient and working SyncML implementation that resides in different places in the specifications.

Chapter 4: SyncML Fundamentals

This chapter lays the ground for the following parts that give in-depth insight into the different components of SyncML. The key design goals such as interoperability, performance, scalability, and footprint are introduced. The chapter further explains how SyncML matches these goals.

It gives an introduction into the overall SyncML architecture and how the different specifications fit together to a working standard.

Chapter 5: Synchronization Protocol

The Synchronization Protocol [SSP02] needs to be studied to understand the SyncML technology and how different SyncML components interact with each other. This chapter analyzes this protocol, which uses the functionality from most of the other SyncML components, e.g., the Representation Protocol [SRP02] or the Meta Information DTD [SMI02].

In addition to the phases and features of the Synchronization Protocol, this chapter brings out many crucial issues, which need to be taken into account when implementing this protocol. These can substantially help to achieve a high-quality implementation of a SyncML Client or a SyncML Server.

Chapter 6: Representation Protocol

The Representation Protocol defines the way a SyncML document, which is exchanged using the Synchronization Protocol, has to be built. It defines the DTD on which the SyncML XML document is based. This chapter describes the different commands and DTD elements as a tutorial, using examples and giving hints and tips not found in the specifications themselves.

Chapter 7: Supportive SyncML Components

This chapter covers the Meta Information DTD and the Device Information DTD [SDI02] specifications, which support the Synchronization Protocol and the Representation Protocol for enabling complete and fully operational SyncML sessions. These specifications complete the SyncML synchronization framework.

In addition to the above, this chapter provides a comprehensive description of the three SyncML transport bindings, which enable the SyncML functionality over HTTP [RFC2616], WSP [WSP01], and OBEX [OBEX99]. By understanding these bindings, SyncML can easily and efficiently be provided in an interoperable manner over the most common transport protocols.

Chapter 8: Security and Authentication

Security is an important aspect in all data processing standards and applications, but especially in cases were data is exchanged using public packet-switched networks, security needs special attention.

This chapter explains the different levels an application using SyncML can secure the exchanged data, it describes SyncML's built-in security features and uses examples to show the different implications in several scenarios. It demonstrates how SyncML security can be enhanced using additional techniques and gives an outlook of possible future enhancements.

Chapter 9: Device Management

The latest major functional addition to the SyncML family of specifications is device management, called SyncML DM [SDM02]. This chapter gives an introduction to the unique area of managing mobile devices and what this exactly means. It explains how SyncML is used to provision and manage mobile devices providing synchronization and other value-added services. It covers the differences with the data synchronization parts of SyncML and what is needed for a SyncML implementation to manage devices.

This uses a number of illustrative examples and scenarios focusing on the different aspects of this new field for SyncML. In addition, the aspects for achieving interoperable device management systems by leveraging the SyncML Device Management Framework are analyzed and discussed in this chapter.

Part III: Building SyncML Applications

This part contains tools, hints, and tips that speed up developing SyncML applications on servers as well as on clients.

Chapter 10: SyncML API and Reference Implementation

SyncML provides a Reference Implementation in the C language to help companies to quickly enable their implementations to support SyncML. This chapter describes the modular architectures and how to optimize the Reference Toolkit for different environments. The chapter also explores the design and API of the Reference Toolkit and uses multiple examples to show how to exploit it efficiently.

Chapter 11: Mobile Devices and SyncML

This chapter describes the special considerations to be made for the efficient use of SyncML in mobile devices. It covers the special characteristics of mobile and wireless devices and how to account for such characteristics to build implementations that perform well, are robust, have small footprints, and interoperate. It explains the design decisions that SyncML chose to enable the development of small and efficient client implementations. It also addresses the issues, which need to be taken into account by the Servers when communicating with the mobile devices.

Chapter 12: The SyncML Server

As with the SyncML Client, there are several ways to implement a SyncML Server. Some are more efficient than others. This chapter introduces the components that constitute a SyncML Server and which design decisions need to be taken to build servers that perform well and scale to a large number of SyncML Clients.

Chapter 13: Interoperability Verification

The success of SyncML is dependant on the ability to synchronize different devices with each other. Customers will use SyncML only if its various implementations are interoperable.

This chapter introduces the SyncML Interoperability Committee, which was created by the SyncML Initiative to develop and coordinate the SyncML Conformance and Interoperability Process. This chapter describes the steps that need to be taken to first obtain the right to use the term "SyncML conformant" and in a second step, get the SyncML Interoperability logo.

This chapter also contains a description of how to use the SyncML Conformance Test Suite and how the SyncFests are organized.

Part IV: Summary and the Future

This part describes the possible future directions of SyncML.

Chapter 14: Summary and Future

This chapter summarizes the content of the book and gives an outlook on future enhancements of SyncML, and applications based on SyncML that might emerge in near future. It describes the missing pieces and what might come in the next version of SyncML.

Part V: Appendices

Appendix A: Bibliography

In this appendix, we reference the sources that we used. In addition, we point to material that is very helpful for learning more about those areas that are not the focus of this book.

Appendix B: Glossary

The glossary briefly explains some of the most important terms.

Appendix C: Trademarks

A list of trademarks referenced in this book is presented.

Index

The index helps to locate terms used and explained in this book.

About the Authors

Uwe Hansmann is currently a development manager at IBM for various Pervasive Computing projects including the development of the SyncML Reference Toolkit. He was co-chair of the SyncML Core Expert Group. He helped to get several industry initiatives started, was Secretary of the Open Services Gateway Initiative as well as the Secretary and a Board member of the OpenCard Consortium. Uwe received a Master of Science from the University of Applied Studies of Stuttgart in 1993 and an MBA from the University of Hagen in 1998. He joined IBM in 1993 as a software developer and led the technical marketing support team for IBM Digital Library in Europe before joining IBM's Pervasive Computing Division in 1998.

In the past he has written books about Smart Card Application Development Using Java, as well as the Pervasive Computing Handbook. He can be reached via email at hansmann@de.ibm.com.

Riku Mettälä is currently working as a technology manager for SyncML based application technologies in Nokia Mobile Phones. He has headed the software projects related to SyncML and was also involved in the standardization of SyncML from its very beginning. Before working with SyncML, he took part in the Bluetooth standardization effort and also acted as the chairman of the synchronization expert group within Bluetooth SIG. He has worked for Nokia Mobile Phones since 1998 and before that, he was a part of the Nokia Networks organization. Riku has a Master of Science in computer science, with specialization in the telecommunications area. He studied in Tampere University of Technology, Finland and in Virginia Tech, USA. He can be reached via email at riku.m.mettala@nokia.com.

Apratim Purakayastha is currently a research manager in the IBM Thomas J. Watson Research Center. For the last few years he has engaged in research in the area of mobile computing and in mobile data synchronization. He is one of the founding contributors to the SyncML standard and was involved in SyncML from its nascent stage as the basic notions in SyncML emerged from a project in IBM/Lotus. He has authored numerous research papers in the area of mobile computing and data synchronization. He has also played key roles in IBM's synchronization related activities. He joined IBM in 1996 after completing his Ph.D. in Computer Science from Duke University, USA. He can be reached via email at apu@us.ibm.com.

Peter Thompson is currently working at Starfish Software as a Senior Software Engineer. He is also the chair of the SyncML Technical Committee and is active in as many of the Expert Groups as he can spare time. Peter is the least educated of the authors with only a Bachelors of Science from the University of California at Berkeley. When not working on SyncML or with other Standards organizations, he also flies his airplane around California. Peter can be reached at peter.thompson@starfish.com.

Acknowledgments

We had a unique opportunity to work with colleagues from leading companies around the world on SyncML, which helped us broaden our experience. Without that we could not have written this book. We would like to thank all members of the SyncML Initiative and all people who have invested an important time of their lives to create SyncML, as well as our employers for having provided us with this opportunity. We would like to thank everybody who worked with us on SyncML, especially Frank Dawson, Douglas Heintzman, Rajiv Jain, Ilari Nurmi, Noel Poore, Timo Saraketo, Dan Wolfson, and Quinton Zondervan.

Numerous people furnished us with in-depth reviews of the book, supported us, or provided us with their invaluable expertise. We are indebted to Chatschik Bisdikian, Stephane Bouet, Douglas Heintzman, Philippe Kahn, Hui Lei, Michael Robert, James Scales, and Thomas Stober.

We would especially like to thank our wives and kids Alisha, Anna, Anusha, Eeva, Linda, Sandra, and Urmi for the borrowed time.

INTRODUCTION

1

An Introduction to Data Synchronization

Mobile computing gives everybody access to their business or personal data everywhere, using devices like Personal Digital Assistants (PDAs), smart phones, mobile phones, and laptops. An online connection to a corporate datastore might not always be possible, due to lack of network coverage, for example. Sometimes even if a connection is available, using it might not necessarily be the fastest and most cost-effective way for the application to operate. In situations such as these, data synchronization is a key technology to alleviate those shortcomings.

Data synchronization allows a consistent local "copy" of various kinds of data, from a central corporate datastore or a service provider datastore on the user's device. It is therefore possible to look up or change data locally on the device without requiring an online connection to the master copy of that datastore. The simplest case is a user retrieving data for his local copy. Here, the application needs to get only the changes from the master datastore to the local copy, without copying the complete datastore again. Synchronization gets more complicated as soon as a lot of different users make modifications to their local copies of the datastore. Now, somehow, these modifications need to be reconciled between all copies of that datastore.

Data synchronization is the technology used to keep all these distributed copies of a datastore consistent by communicating the actual changes between these copies and by resolving conflicts that may arise due to contradictory changes in different copies of the same datastore.

Today, synchronization services support Personal Information Management (PIM) data, such as addresses, calendar entries, memos, and to-do's, as well as Relational Databases and file systems.

The following paragraphs describe the different possible synchronization topologies: one-to-one, many-to-one, many-to-many, and two hybrid versions. These definitions are followed by explanations of the different synchronization modes: local, pass-through, and remote.

The different challenges and problems that arise while keeping data synchronized are elaborated in the following part. This chapter closes with an overview of related standards organizations, most of which have chosen to change their synchronization technology in favor of SyncML®.

The Different Topologies

Changes made to different copies of a datastore can be propagated to other copies of that datastore in different ways. The synchronization topology defines the logical flow of the changes propagating through the network of computers hosting instances of that datastore. The four major topologies are:

- One-to-one
- Many-to-one
- Many-to-many
- Hybrid of many-to-one and many-to-many

One-to-One

The one-to-one topology is the simplest case. The other topologies can be seen as an extension of this one. Here the data is only shared between one server (the square in Figure 1–1) and one client (the circle in Figure 1–1). A possible usage scenario for this topology is a datastore that is mirrored for backup purposes. All changes made to the client are

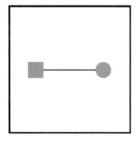

Figure 1–1
One-to-one topology

also sent to the server to ensure that its copy of the data reflects the current version of the client copy. Assuming that data is only changed in the client directly (i.e. no modification is made to the server copy besides synchronizing with the client), then there is no risk of any conflict in this topology. The one-to-one topology is also known as the "Dedicated Pair" topology.

This kind of topology is also used between someone's PDA and personal computer, with the difference that changes are usually made on both the PDA and the personal computer. In this case, the conflicts are typically identified by the PC and directly resolved on the PC. In some cases the conflict is marked and the user is asked to resolve it.

Many-to-One

Numerous commercial systems are examples of the many-to-one topology (also known as central master or star topology). In this topology, data is propagated from a central master to the different entities containing copies of the data, as shown in Figure 1–2.

The main advantage of many-to-one topology is its relative simplicity to implement compared to many-to-many topology, which is described in the next section.

All clients exchange data with the central server only—two clients cannot exchange data directly without the intermediary central server. Because of this characteristic, conflicts can only arise at the central server, which needs to detect and resolve them. The clients themselves do not need to worry about conflicts. They just inform the central master about the local modifications and process the change requests they receive from the central master. There is no need for the client to determine where to send it, as in the many-to-many topology.

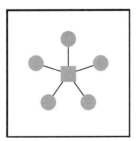

Figure 1–2
Many-to-one topology

This topology is common when a person has a PDA, a cellular phone, and a personal computer sharing an application such as the calendar application, and both the cellular phone and the PDA are synchronized with the personal computer (but not between themselves). This kind of interaction is also common when family members carry cellular phones and update their shared family Web calendar independently or when mobile employees in an enterprise update inventory datastores independently.

The drawback of this architecture is that the central master could become a bottleneck, a single point of failure that could immobilize the entire system. Let's consider an Internet service provider scenario with a central master that serves several hundred thousand accounts, all trying to synchronize with the same central datastore. Here the central master should not be a single server, but a cluster of high-performance servers to limit the latency in response time even if one of the servers fails.

Many-to-Many

In many-to-many (or peer-to-peer) topology, there is no central server. Every client is also a server, as shown in Figure 1–3. For simplicity in this chapter, the client/server combination on each device in the many-to-many topology is just called client.

Every client gets updates from and sends updates to every other client. After a record on one client is updated, this client is responsible for updating all the other copies of the data on all the other clients to ensure that the consistency of the distributed datastore is maintained. This might be by directly contacting the other clients or by sending the updates to the clients nearby, which are then responsible for propagating it further.

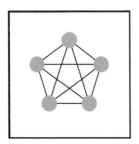

Figure 1–3
Many-to-many topology

Consequently, every client must be able to detect and resolve conflicts. This requires more complex software on each client, which naturally increases the implementation cost, especially on small mobile clients, like mobile phones, in which memory is a scarce resource.

Compared to the many-to-one topology, the many-to-many topology is more robust but also clearly adds to the complexity. In this topology, it is very difficult to find out if a modification was indeed propagated to all clients at a given point in time.

One advantage of the peer-to-peer topology is that without a central server, there is no single point of failure. Every client has a copy of the data and can act as a server. The clients can continue to work and exchange data despite failures in other parts of the network. A client can retrieve updates from the closest server in the network, which gives quicker access to data otherwise stored remotely.

This topology may occur whenever there is no notion of a primary datastore involved in the system. Consider a team of emergency response workers taking readings such as measured temperatures, toxin levels, and structural stress conditions in a building or an affected area. They can synchronize these readings as they pass by each other using direct wireless or infrared links between their handheld devices.

Hybrids of Many-to-One and Many-to-Many

In an effort to combine the advantages of many-to-one and many-to-many topologies, hybrids containing characteristics of both types can be used, as shown in Figure 1–4.

The cluster consists of a two-level structure of data copies. The top level consists of a cluster of servers. All servers contain copies of the data and replicate between each other, but for each data object only one

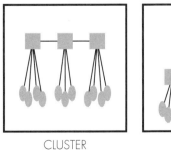
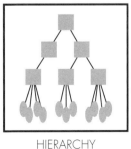

Figure 1–4
Hybrids of many-to-one and many-to-many topologies

server keeps the authoritative copy. The other servers are unaffected by the failure of one of them. Using geographically distributed servers can contribute to reducing the distance between server and clients.

In a hierarchy, the server structure could be modeled according to the organizational structure of a company. The top part of the figure shows servers, which are at the same time clients of a server one level above. In this structure, even when one section experiences a failure, the overall topology can still work properly.

In commercial implementations using a central master topology, the master server itself consists of a cluster of servers accessing a central datastore. This setup guarantees high availability and reduces the disadvantages of a central master topology with regard to the single point of failure. Nevertheless in this setup the servers are physically at the same location and a network failure could make them unreachable. That would not be the case in the cluster or hierarchy topology, as described above.

Summary

Clearly, the one-to-one and the many-to-one topology are abundantly more common in the context of day-to-day commercial applications than the many-to-many topology. Moreover, the many-to-many topology can be indirectly (but inefficiently) achieved by designating one device as a server and stipulating that the other devices synchronize with that server and hence indirectly synchronize with each other via that server. Implementing the many-to-one topology (which includes the one-to-one model) is conceptually simpler and the resulting implementations are orders of magnitude simpler than the ones that support many-to-many topology. In the many-to-many model, complex data structures such as "version vectors" need to be associated with data items to correctly synchronize data. The many-to-many model is also especially stubborn for the purposes of accounting and failure recovery.

For the above reasons, SyncML is optimized for the many-to-one topology. It allows the exchange of datastore sync anchors (see Chapter 5) in the beginning of a synchronization, which indicate the last "timestamp" at which the two computers synchronized. The timestamp could be an actual time value or a logical counter. Based on the exchanged sync anchor values, the associated sync engines could use simple data structures such as change logs (see Chapter 4) to determine the changes made to data items since the last instance of synchronization. SyncML, however, *allows* many-to-many synchronization. It allows each data item to have an associated version which could actually be a version vector required for many-to-many synchronization. It also does not specify the format of the sync anchor explicitly and therefore that could also be a version vector. Furthermore, a SyncML device can play dual roles of a server or a client.

The Different Usage Modes

Synchronization is typically used in the following three physical ways or modes: local, pass-through, and remote. Each synchronization mode has its specific challenges, which are described below.

Local

This is currently the most common way to use synchronization. A typical scenario, shown in Figure 1–5, has a user synchronizing data from the PC or laptop on his desk with a PDA that is connected to the PC or laptop through a serial, Universal Serial Bus (USB)™, infrared, or Bluetooth™ connection.

An application running on the PC acts as the synchronization server. The data is usually retrieved from and stored in another application on the computer, such as Lotus Notes® or Microsoft Outlook®. This scenario requires the user to first synchronize his desktop application with the server (e.g. Lotus Domino™ or Microsoft Exchange®) to retrieve the latest updates from the server before synchronizing the PDA with the desktop.

Pass-Through

This scenario can also be called "router." In this scenario, the synchronization client is connected locally to a computer, similar to local

Figure 1–5
Local synchronization

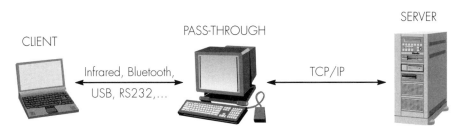

Figure 1–6
Pass-through synchronization

synchronization. But in this usage mode, the local computer is used as a router to pass the synchronization requests to another server acting as the "true" synchronization server. In this usage mode, shown in Figure 1–6, a device can synchronize with a remote server thanks to the pass-through service on the local computer bridging the technologies, even if this device only supports infrared or Bluetooth connectivity.

While in the office, the device can use the faster and cheaper network connection between a desktop computer and the server. A drawback is that the synchronization is now done with a remote server, even while in the office, which generates some additional network traffic compared to the local synchronization case. While on the road, the user uses the remote synchronization to directly synchronize with the synchronization server, as described in the next paragraph.

Remote

Remote synchronization, shown in Figure 1–7, is used if the synchronization client is connected to the synchronization server using a network connection. This might be a local area network (LAN) or a wireless network like GSM (Global System for Mobile Communications) or UMTS (Universal Mobile Telecommunications System).

With wireless networks like GSM, though, the data exchange rate available is obviously not as high as when connected to the synchronization server by a LAN. The advantage of a PDA and/or a mobile phone is that the user can access the server and synchronize the data from almost everywhere in the world.

Figure 1-7
Remote synchronization

Challenges with Data Synchronization

Data synchronization technology is a very important part of enabling offline usage of data, but it also presents some challenges that need to be solved in order to provide a sound solution. The following section focuses on the central master topology, since it is currently the most widely used.

The main challenges are:
- Conflicting updates
- Conflict detection
- Conflict resolution

Another challenge with data synchronization is the need to agree on a common content format for data exchanges. For PIM data, SyncML requires the support of specific content formats to ensure successful synchronizations.

Conflicting Updates

In Data Synchronization, where copies of the same data reside on more than one device, it is very likely that the same entry is manipulated on more than one device.

Consider a datastore with customer records, containing the customer's name, address (street and city), and data about recent orders. This data is centrally stored in the corporation's mainframe computer. The salesmen who visit customers during the day synchronize their PDAs with this corporate datastore every morning and every evening.

During his visit at a customer, the salesman changed the fax number for the customer "Smith" to "12345". The same day, the customer relations department received a letter from Mr. Smith notifying them that his fax number has changed. The customer relations agent updated

Mr. Smith's record, but unfortunately made a typo while entering the number and updated the entry with "92345".

Two conflicting updates to the same record have been made since last time the salesman synchronized the customer datastore. It is now up to the synchronization server to first detect the conflict and then resolve it.

If the synchronization server could not detect the conflict, it would send the update it had received from the customer relationship department to the salesman's PDA and would also update the master datastore with the update the salesman made during the day. As a result, the content of the two copies would not contain the same data–the master datastore would have the fax number "12345", and the salesman's PDA would contain "92345". This problem would probably remain, undetected and uncorrected, until somebody updated the same object again. The problem described below is illustrated in Figure 1–8.

The update conflict described above is known as a *write-write conflict*: The same field is updated in the same time period in two different

Figure 1–8
Synchronization without conflict detection

copies of the datastore. Other update conflicts include *read-write conflict* and *constraint violations*. They are explained later in this chapter.

For example, examine the following scenario: During the day a salesman is entering a new order for the customer with the name "Smith" and changes his fax number from "98765" to "12345". At the same time, the corporate marketing department is creating a list of all customers' fax numbers for a mass mailing. The salesman hasn't synchronized his PDA yet, and therefore the central datastore's entry for the fax number of the customer "Smith" still contains "98765", where "12345" is the correct value. This situation is called a *stale read*, where one retrieves an entry from one copy of the datastore that is not up to date because a change was made to another copy of the same datastore and wasn't propagated to all other copies of the datastore. Figure 1–9 illustrates this situation.

In the case of a read-write conflict, the update to a data field is made based on the value of another field in the datastore. While making this update in one copy of the datastore, the field that determined the decision to update the other field was changed in another copy of the datastore. Therefore the update is made based on a stale read. Detecting these conflicts is especially important for Relational Databases. PIM systems rarely have these conflicts.

One possible example of a conflict based on a constraint violation is where a certain field in the datastore is constrained to accept only values

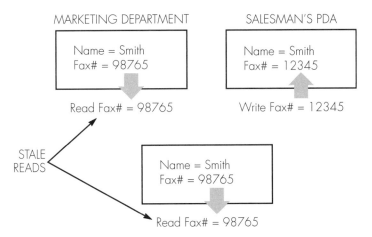

Figure 1–9
Stale reads

between "1" and "10". Now this constraint is changed such that only values between "1" and "5" are acceptable. In another copy of the datastore, somebody enters a value of "8" into the field with this constraint. During the next synchronization the synchronization server has to detect this conflict if the datastore is to be kept consistent.

One possible way of preventing conflicting updates (especially read-write conflicts) is to obtain exclusive access to an object before updating it. In the central master scenario this would require the device to contact the central master and request exclusive access. This might be possible in intranet scenarios, but is not an option with mobile devices. Mobile devices might be in areas that currently have no network coverage. Also, setting up a wireless network connection, for example via GSM, takes a few seconds and can be quite expensive (especially when roaming internationally). Therefore getting exclusive access to the object before updating it is not a practical alternative. Thus, a synchronization server must be capable of detecting and resolving conflicts.

The conflicts mentioned above all belong to a category of conflicts called *mechanical* conflicts, since there have been concurrent modifications to the same record. It is also possible, however, for a user to enter a meeting with a customer from 8:00 to 10:00 in the morning while his secretary blocks the 9:00 to 10:30 slot for a meeting between the user and his manager. From a synchronization technology point of view, there is no conflict. Both records were newly added distinct calendar entries. But the user now has to attend two meetings from 9:00 to 10:00, which is a *semantic* conflict that the user needs to solve. The calendar application can in this case support the user by detecting the conflict and informing the user about this.

Conflict Detection

Conflict detection is important in order to keep different datastores in "sync," meaning keeping them consistent.

One important prerequisite of being able to synchronize data is the ability to uniquely identify each record in the datastore. Relational Databases have primary keys that serve this very purpose. In other types of datastores, such as PIMs for example, UIDs (unique identifiers) are used. The UIDs in the datastore on the central master are usually called GUIDs (globally unique identifiers; possibly only unique in the scope of that one datastore). Each copy of the datastore on a client has

Challenges with Data Synchronization

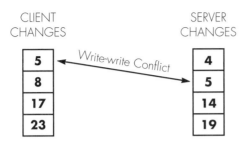

Figure 1–10
Identifying an update conflict

its own UIDs to identify the records, which are called LUIDs (locally unique identifiers).

The central master maintains a datastore to store the LUID of each record for every client that synchronized a given datastore. While synchronizing, the central master uses this LUID to tell the client which record to modify.

A synchronization session usually starts with the client sending a list of changed records since the last synchronization for a particular datastore to the server. Next, the central master then generates a list of all modifications that occurred during in the same time period. To detect possible conflicts, the server then compares the two lists and identifies every LUID/GUID combination that exists in both of the lists as a conflict, as shown in Figure 1–10.

Conflict Resolution

Conflict resolution is the action that the server must take after a conflict has been detected.

There exists a wide variety of methods a server can use to resolve conflicts. The easiest way for the server is to duplicate the entries and to mark them as conflicts. This strategy does not lose any data, but puts the burden of resolving the conflicts on the user (who has to manually choose the correct record or merge the two records).

The next possibility is for the server to identify one of the two conflicting records as the winner and delete the other one. The decision can be based on user preferences, such as:

- Updates made on the client always win.
- Updates made on the server always win.
- The latest change wins.

Basing the decision on where the change was made is a pretty simple method to implement. Making the decision based on which one of the two conflicting updates was the latest one is more complicated. First, this requires the datastores to record the time when the change was made. This is often impossible, especially with mobile devices, like mobile phones, or PDAs, like the Palm Pilot™. For this to work, the system time needs to be synchronized across all the devices involved, or a scheme must be set up to address time-drift.

The third method to resolve conflicts is to merge the two conflicting records into one new one. This could be done in cases where complete records are synchronized, for example an address book record containing name, street, city, and phone number. Additionally, this requires the synchronization server to understand the structure of the data that is synchronized and to be able to identify the field within that structure that was changed. This is not possible if the records contain only single fields that are to be synchronized, and the synchronization server cannot interpret the data in the field.

Resolving conflicts with PIM data is relatively easy. It is more complex with Relational Databases, which usually have relationships between different records that need to be honored during conflict resolution.

Related Work

Synchronization is an important component of many standards for mobile and wireless technology. The SyncML Initiative is closely working with other organizations, especially Infrared Data Association (IrDA®), Wireless Application Protocol (WAP) Forum™, and Third Generation Partnership Program (3GPP)™, to get them to agree that SyncML (Data Synchronization and Device Management) should be the synchronization technology of choice in their specifications.

Infrared Mobile Communication

The IrDA mobile communications committee defined the Specification for Infrared Mobile Communications (IrMC) [IrMC00], to provide information exchange over infrared. IrMC defines five levels of information exchange:

- Level 1 (Minimum Level)
- Level 2 (Access Level)
- Level 3 (Index Level)

Related Work 17

- Level 4 (Sync Level)
- Level 5 (SyncML Level)

Level 1 to Level 4 of the IrMC specification define the exchange of a limited number of different objects, such as business cards, calendar data, messages, and email data over personal area networks with connection-oriented or connectionless links. They do not support the synchronization of other forms of data such as Relational Databases or tabular data. Since the IrMC specification was initially designed for local data synchronization, the methods proposed are not optimized for data synchronization over wide area networks, such as synchronizing the phone book on a mobile phone with one of the corporate public address book datastores over the Internet.

Level 5 was added in 2001 and defines SyncML as the synchronization technology used in IrMC. The SyncML Initiative has closely worked with IrMC in order to produce a Synchronization Profile that strongly recommends SyncML as the preferred synchronization technology.

Wireless Application Protocol

In June 1997, Ericsson®, Motorola®, Nokia®, and Unwired Planet® (now Openwave®) founded the Wireless Application Protocol (WAP) Forum as an industry group for the purpose of extending existing Internet standards for the use of wireless communication. By spring 2002, the WAP Forum® had more than 500 member companies from all parts of the industry, including network operators, device manufacturers, service providers, and software vendors.

The WAP Specification Version 1.1 was released in the summer of 1999, and the first WAP devices and services were available as early as the fourth quarter of 1999. Version 2.0 of the WAP Specifications was approved and released to the public in February 2001.

Since version 2.0, WAP requires all WAP Servers and WAP Client Devices supporting data synchronization to use SyncML. WAP also requires clients and servers to pass the SyncML Conformance testing as a prerequisite for passing the WAP Conformance testing.

The Open Alliance® was formed in June 2002. OMA was created by the consolidation of the supporters of the Open Mobile Architecture® initiative and the WAP Forum. Additionally, SyncML Initiative® Location Interoperability Forum (LIF), MMS Interoperability Group (MMS–IOP), and Wireless Village® announced they had signed

Memorandums of Understanding of their intent to consolidate with OMA. This consolidation should take place by the end of 2002.

Third Generation Partnership Program

The Third Generation Partnership Program (3GPP) was established in December 1998 as an organization of all partners interested in the evolution of mobile systems to the third generation evolving from the GSM technology. GSM is looked at as the second generation, GPRS (GSM Packet Radio Service) and EDGE (Enhanced Data Rates for GSM Evolution) as the 2.5 generation, and UMTS, UTRA (UMTS Terrestrial Radio Access), W-CDMA (Wideband Code-Division Multiple Access), and FOMA (Freedom Of Mobile multimedia Access) as the third generation of mobile communication technology.

SyncML and 3GPP are working together closely, and SyncML technology has been mandated as the method of choice for data synchronization since Release 4 of the 3GPP specifications.

Part I
SyncML Overview

2

SyncML: An Introduction

In the past few years we have observed a sustained growth in the use of mobile computers. Small handheld devices and mobile phones with data communication capabilities are already common. Mobile Personal Information Management (PIM) applications, such as calendars and address books, are commonplace and mobile business applications, such as inventory control, are emerging. For many mobile applications, the data on a handheld device corresponds to the data on a personal computer or a network server. As Chapter 1 explains, this data must be kept consistent using data synchronization. The basic process involved in this kind of data synchronization, including fundamental algorithms and data structures, is well known in the mobile computing community.

Synchronizing a handheld computer or a mobile phone with a personal computer is already popular. Applications on the Palm® and the PocketPC™ computers are synchronized with their counterparts on the personal computer. Such *local synchronization* is typically performed using serial communication over a cable or an infrared connection between the handheld device and the personal computer. Usually the same vendor writes the applications for the handheld device and the personal computer. The vendors also control the associated data synchronization protocols and data formats. This development model works fine in practice, because the set of applications is primarily limited to PIM applications, the data is associated with a single individual, the communication mechanism is predominantly serial, and the number of commercial handheld platforms is relatively small.

Advances in wireless data communication are making it possible for wireless-enabled handheld computers and high-end mobile phones to access network services. A model is emerging where the counterpart for a calendar application on a handheld device is not a desktop application but a network service offered by a Web portal or an enterprise server. This model is attractive for a number of reasons. First, the network service provider relieves the user of managing application data on his or her desktop. Second, the handheld application can interact with its network counterpart more spontaneously without the need for physical proximity to a personal computer. Third, the model enables the sharing of application data across users to build useful applications such as family calendars. Finally, the model facilitates business applications on a variety of handheld devices and mobile phones by allowing the integration of data on handheld devices with business data in back-end datastores.

Despite its relative advantages, the above model is far from being truly realized. The model requires *remote synchronization* of application data on handheld devices and mobile phones with data in various datastores in network servers. Unlike local synchronization, it is not feasible for application vendors to fully control the synchronization process. One reason is that currently many mobile phones do not expose application-programming interfaces. Another reason is that network service providers and handheld device manufacturers are usually from distinct business segments, and it is difficult for them to cooperate on this synchronization. It is still conceivable that a substantially sizeable application vendor such as Lotus® or Microsoft® would want to write synchronizing client applications for Lotus Notes® and Microsoft Exchange® for handheld platforms like PocketPC that publish open programming interfaces. This approach has only had limited success, as the application vendors could only support typically one or two platforms and very few specific communication protocols.

A number of synchronization vendors have emerged to "bridge the difference" between server applications and handheld or mobile phone applications. The vendors typically partner with a client platform provider and a network service provider or a server application vendor. They also often make a choice of the kind of communication protocol to support. Thus, they typically offer a "synchronization solution" that works for a set of clients and set of server applications or service providers. They also typically use proprietary synchronization protocols and data formats that are optimized for the set of platforms and supported applications without concern for interoperability. This has resulted in disjointed islands of communication.

Figure 2–1 illustrates the segmentation of the mobile data synchronization market in the period from the middle of 1999 through 2000. The figure primarily focuses on remote synchronization capabilities and not on mobile device to personal computer synchronization capabilities. On the left, examples of mobile devices are shown. On the right, some examples of Web portals and server applications are shown. In the middle, some synchronization vendor products are shown. It is clear that for mobile phones not much remote synchronization capability with Web portals and enterprise applications existed in the indicated timeframe. FoneSync™ only offered synchronization between certain mobile phones and personal computers. TrueSync® offered synchronization between Motorola® phones and a few portals such as Excite@Home™ and Yahoo!®, and to Microsoft Exchange. Intellisync Anywhere® offered synchronization between the Palm and Symbian

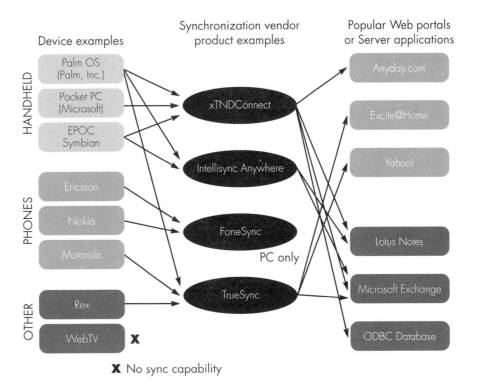

X No sync capability

Figure 2–1
The figure illustrates the pre-SyncML® segmented mobile data synchronization market from mid-1999 through 2000. This figure is primarily for illustrative purposes and may not accurately reflect the exact features of certain products in the indicated timeframe.

OS®-based handhelds and Lotus Notes and Microsoft Exchange. The xTndConnect® product offered synchronization between Palm, PocketPC, and Symbian® devices and Anyday.com™ portal and enterprise PIM applications, such as Lotus Notes, Microsoft Exchange, and Open Database Connectivity (ODBC) [San98] compliant databases. Many synchronization vendors did not offer a direct remote synchronization solution from the mobile device to the server but used the personal computer as a pass-through to the network server, thereby limiting many cited benefits for true remote synchronization. Now consider, if one installed the TrueSync client on a Motorola phone to synchronize with Excite, that person would need to find a vendor application that synchronized the Motorola phone with the Lotus Notes application and install it on the mobile phone. From the perspective of the consumer, if one has a client application that one wants to synchronize with a server application, one must first find a supporting synchronization vendor and install the required vendor-specific client code. If one now wants to synchronize a different application, one may have to find a different synchronization vendor. From the perspective of a network service provider, the number of sync vendors it is able to partner with defines its reach of client platforms.

The SyncML Initiative

The market segmentation referenced above constrains consumer freedom of choice of platforms, applications, and service providers. The segmentation also constrains the reach of network service providers and server application vendors with respect to the number of clients they can support. Overall it limits the growth of mobile applications, as data synchronization is a key requirement of many mobile applications. The key purpose of SyncML is to define an open specification for data synchronization such that client and server applications can be developed independently. The SyncML Initiative targets the specifications to become the de facto standard for data synchronization through wide adoption of the open specifications and numerous conforming implementations. Applications on clients and servers that support SyncML and use SyncML conformant data formats will be able to synchronize with each other. In the ideal vision of SyncML, applications on any phone or handheld device can synchronize with corresponding applications on a server platform or another device. SyncML is primarily targeted and designed for remote synchronization between a mobile client device and a server, but can also be used for local synchronization and synchronization between networked devices.

The reader may be interested in a brief history of the SyncML Initiative. During early 1999, IBM® and Lotus were exploring ways to enable mobile applications that synchronized data with IBM databases and Lotus Notes. Clearly, as a provider of infrastructure software, IBM/Lotus wanted the ability to do the above in a way that worked with a large number of mobile clients, irrespective of the specific nature of the client. They strongly felt the need for an open synchronization standard. At about the same time, companies such as Nokia® and Ericsson® were exploring the issue of data synchronization in the context of Infrared Mobile Communications (IrMC) [IrMC00] and the Bluetooth™ Special Interest Group (SIG)[MB01]. Motorola/Starfish® was considering opening up its proprietary synchronization protocol. IBM/Lotus and Nokia together took the early lead in developing the SyncML Initiative and actively evangelized it in the software and telecommunication communities. With a draft of the specifications and substantial work performed in informal cross-company working groups, the SyncML Initiative was founded in February 2000 with Ericsson, IBM, Lotus, Motorola, Palm, Psion®, and Starfish as the founding sponsors. SyncML gathered momentum at a steady pace during 2000. The first version of the specification was released in December 2000 with a supporting reference implementation. In July 2001 the organization transitioned to an incorporated nonprofit entity to better respond to the needs of the growing data synchronization community. At the same time, Matsushita®, Openwave®, and Symbian joined as sponsoring members of the SyncML Initiative. In 2001, the SyncML Initiative enjoyed over 600 supporter companies, some of which are developing SyncML compliant products.

There are two other kinds of SyncML memberships besides *Sponsor–Promoter* and *Supporter*. Becoming a Promoter member gives companies the ability to participate in the technical activities of the SyncML Initiative. Benefits of Promoter membership include input into the evolution of the specifications through the SyncML Initiative Technical Expert Groups, access to specifications and toolkits under development, early access to documents, tools, and specifications in advance of release to Supporters, and free license to the SyncML Test Tool.

Becoming a Supporter allows early access to review the specification and other related documentation. Further, Supporter membership gives one an opportunity to provide contribution and comments to the specification work and to participate in online forums where Supporter input is valued and welcomed.

An Overview of SyncML

Logically, synchronization between two applications requires the sharing of changes that the applications have made to data common to both applications. Synchronization also involves potential reconciliation of conflicting changes made concurrently. If applications can *represent* what has changed in a manner that is mutually understood, and are able to *communicate* those changes in an agreed upon fashion, they can synchronize their data. SyncML is primarily based upon this observation. Naturally, the two fundamental parts of the SyncML data synchronization specification are the SyncML Representation Protocol and the SyncML Synchronization Protocol. The Representation Protocol is essentially the syntax for specifying the changes that an application has made to its data. The Synchronization Protocol is the specification of the sequence of packages that applications must exchange in order to communicate their changes to each other.

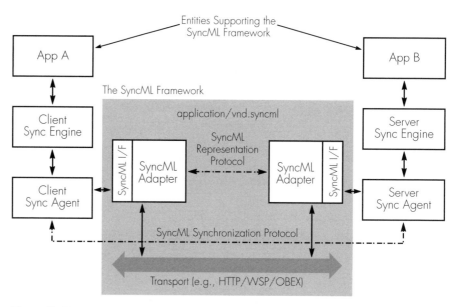

Figure 2-2
The main elements of the SyncML Framework and the entities that support the SyncML framework.

Figure 2-2 illustrates the scope of the SyncML framework. The framework encompasses the Representation Protocol, the Synchronization Protocol, and the various transport bindings such as Hyper Text Transfer Protocol (HTTP) [RFC2616] and Object Exchange Protocol (OBEX™) [OBEX99]. The framework also includes elements of the SyncML reference implementation, such as the SyncML API and the SyncML adapter. The SyncML framework, however, does not include other synchronization-related programs, such as application, synchronization engine, and synchronization agent, that drive the framework. Instead, the framework assumes their existence.

In the figure, application A on a client device wants to synchronize its data with application B on a server computer. There is a logical entity called the "synchronization engine" on both the client and the server sides. The synchronization engine on the client typically only keeps track of local changes that the application makes to its data. The synchronization engine on the server also keeps track of local changes made by the server application, but in addition, it encapsulates intelligence about the semantics of application data. It stores and manipulates supporting metadata called "version" to account for changes made to data by multiple clients. It has methods to detect and resolve what constitutes a conflict between local changes to data and changes obtained from the synchronizing client. Although the synchronization engine plays a critical role in the synchronization process, it is not part of the SyncML specification. The synchronization engine can sometimes be part of the application itself. It is expected that competing mobile applications and synchronization vendors would want to differentiate themselves by advancing their synchronization engine technology.

The synchronization engine is expected to co-ordinate synchronization using a "synchronization agent." The synchronization agent is primarily responsible for generating and processing SyncML Packages in the format and sequence specified by the SyncML Representation and Synchronization Protocols, respectively. The synchronization agent is also not part of the SyncML specification. Thus, although the synchronization agent is functionally defined, the particular ways of implementing the synchronization agent are not specified. The interaction between the synchronization engine and the synchronization agent is also unspecified. In some cases, the application, the synchronization engine, and the synchronization agent may all be combined.

SyncML Basics

One core element of the SyncML specifications is the Representation Protocol. The Representation Protocol specifies the logical structure and format of various SyncML *Messages*. The structure and format of a Message is specified using XML (eXtensible Markup Language). The use of a structured description language is essential and XML is a natural choice given its widespread acceptance and use. Figure 2–3 outlines a high-level structure of a SyncML Message. During synchronization, the synchronizing entities logically exchange *Packages*. Physically, each Package can be partitioned into multiple actual Messages that are communicated. There may be many instances in which a logical Package is broken down into several Messages for communication. One motivation for such Package fragmentation could be the lack of communication buffer space in a resource-constrained client, such as a mobile phone. Another motivation could be that smaller Messages are transmitted with higher degrees of success over a low-reliability and low-bandwidth wireless connection. The reader should associate Packages with logical units of communication during synchronization and Messages with physical units of communication during synchronization.

Each SyncML message is a well-formed[1] XML document. The precise definition of a SyncML message is described in a XML Docu-

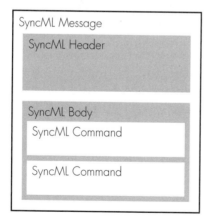

Figure 2–3
The figure shows the logical structure of a SyncML Message

---

1. A well-formed XML document is one that is syntactically correct and conforms to its definition in a corresponding DTD.

ment Type Definition (DTD). The message consists of a header and a body. The header includes information such as the source and target application, routing, session, and authentication. The body of a SyncML message is a set of SyncML "commands." Each command identifies an operation that has been performed or being requested on a data item or a set of data items. Examples of operations include Add, Delete, Replace, Search, and Status. Each command identifies the relevant data items within itself. For example, the Replace command identifies data items replaced by an application in its datastore. When the synchronizing counterpart receives the message and processes the Replace command, it attempts to replace the same items in the corresponding application datastore. If there are no conflicts or errors, the replaced items are now reflected in both datastores. In other words, the replacements are successfully *synchronized*. The Representation Protocol does not impose any implicit order or other constraints concerning how multiple commands in a single message may be processed. For example, if a message contains multiple Add and Replace commands, the recipient may process those commands in any order. The Representation Protocol, however, allows sets of operations to be grouped together with special semantics. For example, the Atomic command element can contain other commands, such as Add or Replace, with semantics that require the recipient to either successfully process all commands within an Atomic group or process none at all. These types of capabilities allow SyncML to be used in commercial applications that may require transactional properties. Chapter 6 offers a more in-depth discussion of the SyncML Representation Protocol.

The SyncML Representation Protocol only specifies the language spoken by two computers when they synchronize. The Synchronization Protocol specifies what the computers should say to each other in what sequence. Consider two persons who know the English language well enough to form correct, meaningful sentences. They will still not be able to have a successful conversation if they do not have an implicit understanding of what sentences should follow what sentences. For example, if two persons alternately say "What is your name?" "I live in London." "I asked you for your name." "London is in England."; the dialogue does not represent a successful conversation, even though both formed correct, meaningful sentences. Similarly, two computers cannot synchronize if they do not follow a prescribed sequence of packages, even if the individual packages are correct. This sequence of packages is specified by the SyncML Synchronization Protocol.

Figure 2–4
This figure is a high-level illustration of the SyncML Synchronization Protocol.

The Synchronization Protocol defines the roles of the client and the server (Figure 2–4). Typically, the client is a mobile device, and the server is a network server or a personal computer. The distinction between a client and server, however, is logical. It is possible for two mobile devices to synchronize using SyncML where one assumes the role of a client and the other the role of a server. Usually, a client requests synchronization in its first message to the server. The server, however, can alert some clients to begin synchronization. For example, a network server can alert a mobile phone client to begin synchronization. The form of synchronization could be two-way or one-way. In two-way synchronization, the client and the server exchange and reconcile their respective updates. In one-way synchronization, only the client or the server sends its updates and the other receives and reconciles them. Two-way synchronization is useful for applications that update shared data concurrently, such as a family calendar or a store inventory. One-way synchronization can be very useful for clients that only need to read the latest data from a server or update a server with the latest data, such as downloading insurance rate quotes or uploading receipt delivery signatures. Chapter 5 provides an in-depth discussion of the SyncML Synchronization Protocol.

Additional Key Considerations

In addition to the Representation and the Synchronization Protocols, SyncML takes additional considerations into account to enable practical, commercial data synchronization. The additional considerations pertain to network binding, acceptable and expected data formats, device capabilities, and security. Some of the additional considerations are embodied in actual supporting specifications, such as device and meta-information specifications. Some other considerations, such as security, manifest themselves inside the Synchronization and Representation Protocols.

Two synchronizing entities must communicate over an actual network. Applications typically communicate using network transport protocols, such as HTTP. Although the SyncML Representation and the Synchronization Protocols are transport-independent, during synchronization the SyncML packages must be carried in actual messages over an actual transport protocol. Clearly, for different transports, the form of these messages could be different, even if they carry the same overall package. The form of messages for a particular transport is called a "transport binding." SyncML defines a few popular transport bindings such as HTTP, Wireless Session Protocol (WSP)[2] [WSP01], and OBEX.[3] The bindings are ancillary specifications that can be used by application developers. If one wants to use SyncML over a transport other than the three above, it is possible to define bindings for additional transports. For example, if one wants to synchronize using an email protocol, an additional transport binding such as Simple Mail Transfer Protocol (SMTP) [RFC2821] can easily be defined. Other potential significant transport bindings could include MQSeries®[4] [GS96], for corporate messaging applications that may want to use SyncML, or Remote Method Invocation (RMI)[5] [Gro01], for applications that may want to keep Java objects synchronized using SyncML.

The application data formats are not specified in the Representation or the Synchronization Protocols. Two applications that synchronize must, however, understand the application data type and data format that they use. SyncML specifications cannot include all possible data formats for all possible data that may be synchronized using SyncML. Consequently, SyncML only identifies a few data formats that SyncML Servers must support for specific content types being synchronized to claim SyncML conformance and test interoperability between Clients and Servers.[6] These data formats relate to PIM applications and are adopted from emerging standards in calendar and contact data formats, such as vCard [VCARD21] and vCalendar [VCAL]. It is

---

2. WSP is a transport protocol specified as part of the WAP communication standard.
3. OBEX is a transport protocol implemented on the Infrared and Bluetooth communication media.
4. MQSeries is an IBM software product that supports messaging between applications running on multiple hosts.
5. RMI is a Java specification that supports object-level communication between applications running on multiple hosts.
6. From this point in the book onwards, the first letters of the words client and server are capitalized to mean a SyncML Client and a SyncML Server.

expected that representation standards will emerge for several classes of business-related data. For example, it is likely that an XML representation for data in Relational Databases will emerge in the near future. It is SyncML's goal to leverage such data standards and incorporate their use into SyncML. It may be useful to think about this process as establishing "data bindings" as such become necessary and appropriate.

It is important for synchronizing computers to know about their respective capabilities, such as supported applications, data formats, and memory limitations. For example, before a server computer sends data to a mobile client, it may be important to know about the client's memory limitations so that the data does not cause a memory overflow in the client device. The synchronizing computers may need to know if they recognize and support each other's data types. The computers may need to know each other's version information for incremental synchronization since the last occurrence of synchronization. They may need to know the maximum allowable size of data identifiers. Datastores on clients and servers are rarely homogeneous. Hence, clients and servers typically use different types of identifiers to identify the same data item. Clients typically use a compact form of identifiers and servers use a more elaborate form. SyncML specifies how this type of device information and metainformation should be represented. Device information and other metainformation are exchanged at the beginning of synchronization between a Client and a Server when they are synchronizing for the first time or if there are changes to device information and meta-information since the last time they synchronized.

Data security is also of critical importance in synchronization. Accessing and updating data must require appropriate authentication. SyncML allows multiple levels of sophistication of authentication. One may choose simple username-password schemes for personal applications, such as a Web calendar. One may choose Message Digest (MD5) based credentials or X.509 Certificates for corporate applications, such as inventory control. In addition, SyncML relies on secure transport layer mechanisms to ensure data protection for sensitive data when in transit across a network.

Another key driving goal of SyncML is the rapid development of mobile applications and prototypes in an interoperable manner. To promote that goal, SyncML also provides a reference implementation of key parts of the SyncML framework on a number of platforms. The reference implementation serves as key enabling code that application writers can readily use to write SyncML applications. It implements the SyncML C-language API and a few transport bindings. The synchronization agent in Figure 2–2 may use the SyncML C API to generate, process, and communicate SyncML Messages. The SyncML C API is not part of the core specification but only a de facto available interface. Applications are free to generate SyncML Messages directly using their own means. The readily available reference implementation has, however, played a key role so far in the acceptance and popularity of the specification.

From an Initiative to a De Facto Standard

SyncML is intended to be a primary mobile application enabler beyond simple Internet access from mobile devices. It is intended to enable data synchronization between a diverse set of computing devices, including mobile handheld devices, mobile phones, personal computers, and network servers. It is designed to be transport agnostic and extensible such that it can work with a variety of network transports. It is designed to accommodate emerging data standards. It is anticipated to span application domains from personal information management applications to mobile business applications.

One key element of SyncML is that it is a message-based specification. Message-based specifications have been successful in client-server and distributed systems. Standards and technologies, such as Component Object Request Broker (CORBA) [Bol01], RMI, and Distributed Component Object Model (DCOM) [Roc98], that depend on acceptance of a particular programming interface have not been widely adopted beyond limited domains. Since SyncML is message-based, it allows platforms such as mobile phones to maintain their proprietary nature and still interoperate with diverse server computers and diverse applications. Another key element of SyncML is the conscious allowance for differentiation. SyncML does not attempt to standardize the functions of synchronization engines that capture diverse application semantics, realizing that it is harmful for application growth and is ultimately not a tractable problem.

The SyncML testing and conformance programs and regular SyncFest events have also played a key role in the proliferation of SyncML. The conformance tests give application writers a basic bar to clear before they are allowed to join interoperability testing with other application vendors at a SyncFest. When an application vendor or a device manufacturer demonstrates interoperability with two or more other entities, the vendor receives the right to use the SyncML Logo.

The logo program has created substantial enthusiasm around SyncML, and efforts to obtain the logo have naturally resulted in more development using SyncML. Chapter 13 elaborates on the SyncML conformance and interoperability certification process.

The price of interoperability is often performance. If SyncML manages to be the lingua franca of data synchronization, but it takes too long to synchronize using SyncML or its implementation stack cannot be accommodated in resource-constrained mobile devices, the industry will not embrace it as a de facto standard. It must be a practical specification. Chapter 4 outlines the various design tradeoffs SyncML makes to realize a practical, useful means of data synchronization.

SyncML targets multiple market segments. The next chapter illustrates representative applications in detail in the consumer and enterprise space. Currently mass market applications like contacts and calendars are primarily beginning to use SyncML. Applications in numerous vertical segments, including banking, insurance, health care, and retail can also use SyncML. Adoption of SyncML in various industry segments is a key to it becoming a universal data synchronization standard.

3

SyncML Applications

SyncML® can enable numerous applications that require data to be synchronized among various devices. The applications range from managing personal information, such as contacts, calendars, and email, to managing enterprise information, such as inventory data. The applications must support diverse devices that connect using different networks. In addition, the applications have varying reliability, performance, and security needs. Authoring such applications and making them interoperable is a difficult task. This chapter takes a closer look at a few applications and explains how SyncML is beneficial for the different entities that use it.

Before considering specific applications, it is useful to note certain characteristics of many common data synchronization applications. Unlike applications that primarily reside and operate on one computer, many data synchronization applications are *partitioned* between Client and Server parts. These two parts of the application *work together* to provide the overall user experience. A Web browser and a Web server also work together to provide an overall user experience. The interaction between the parts of a data synchronization application, however, is more coordinated and semantically coupled than the more ad hoc interactions between a Web browser and a Web server. For example, the address book on a personal handheld device and its PC counterpart are more aware of each other and work in a more tightly coordinated fashion than, say, an MP3 player on a mobile phone and a MP3 download Web server.

Figure 3–1 shows a logical view of mobile data synchronization applications partitioned between clients and servers. Client parts often

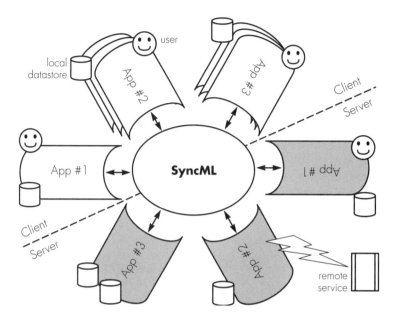

Figure 3-1
A logical view of mobile data synchronization applications showing the partitioning between client and server parts and various external entities that such applications may interact with, such as a user, a datastore, and a remote service.

manage user interaction, keep a record of changes made to application data, and interact with local datastores. Sometimes, the client part uniquely corresponds to one server part as shown in Application 1. An example of this is the address book application of a personal handheld device and its PC counterpart. Sometimes multiple client parts logically correspond to one server part as in Application 2. An example of such an association is found in the family Web calendar example illustrated later in this chapter. The different client parts of the *same* application may reside on different devices, as in the Web calendar application. Those different parts can use various means of wired and wireless communication.

The Server parts of applications interact primarily with back-end datastores and also detect and reconcile conflicting updates. The Server part sometimes also tracks changes made to back-end datastores. The Server part can sometimes interact with a user, as shown in Application 1. For example, the Server part of the address book application may reside on a PC, where a user may directly update address book entries. The Server part, as shown in Application 2, can sometimes implement application-specific logic, resulting in interaction with remote datastores, processes, or services. An instance of such behavior is found in the family Web calendar example below. Depending on application semantics, a Server part may synchronize a single Client datastore with multiple back-end datastores, as shown in Application 3. An instance of such behavior is found in the visiting nurse scenario below. Synchronization Servers typically host the Server parts of many applications. Therefore, pieces of logic common to many applications, such as detection and resolution of certain conflicts, are sometimes factored into a *sync server engine*. For the purposes of this discussion, it suffices to assume that the Server parts subsume functionality often factored into sync engines.

It is natural to ponder at this point what role SyncML plays in enabling these partitioned data synchronization applications. After all, many legacy data synchronization applications, such as Lotus Notes®, are built in a tightly coupled, partitioned fashion. SyncML enables uniform logical communication of data and changes made to data between the Client and Server parts of applications. As a result, the Client and Server parts of applications can be built relatively independently.

SyncML is an enabler for many mass-market applications, such as calendars, email, data backup, and picture galleries. SyncML can also be used for many enterprise applications, such as inventory management, claims processing, and procurement. The two examples below elaborate on the many common applications used today. The applications, however, illustrate the various benefits of SyncML in enabling complete data synchronization solutions. The examples we describe are:

- A *consumer* example: "Coordinating a busy family"
- An *enterprise* example: "Supporting Roving Nightingales"

Coordinating a Busy Family

The Stetsons are a family of four. David, the father, is a sales professional. He often visits clients out of the office. Mary, the mother, works from home. She processes claims for an insurance company. Susan, the sixteen-year old daughter, attends high school. She is also an aspiring ballerina. Mark, the seventeen-year old son, also attends high school. He is an aspiring soccer player. While all of them are busy with their work and activities, they remain a close-knit family. They often schedule activities that involve two or more family members. Such activities include family dinners, going together to the theater, cheering at Mark's soccer games, and attending Susan's ballet performances. The family uses a Web-based calendar service provided by a service provider to schedule several family activities.

Application Setting

Figure 3–2 shows the setting for the Web-based family calendar management application. Different family members use the Client parts of the same calendar application implemented on different devices. Mary uses her home PC to view and make entries in the family calendar. She synchronizes her calendar with the Server's calendar when she dials in to the network. Mark uses a PDA. He uses the PDA's calendar to view his schedule and make changes, and occasionally synchronizes with the Server using the PDA's wireless dial-up connection. David and Susan both use a mobile phone. They operate in the same mode as Mary and Mark but with two key distinctions. They do not have to use dial-up, as

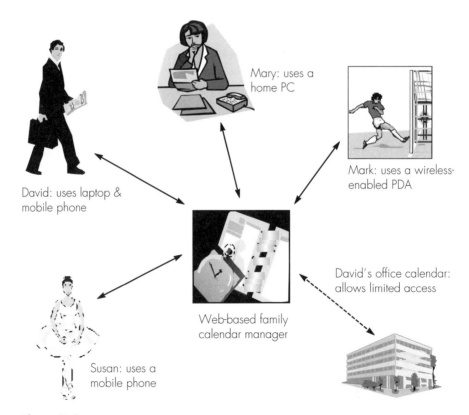

Figure 3–2
The setting for a Web-based family calendar management application.

the mobile phones are mostly connected. They also have the additional capability of being alerted by the Server's calendar. David also uses a laptop version of the family calendar application.

The laptop and the PC versions of the application additionally allow management of application preferences using a Web browser. Clearly, although the different Client calendars operate on the *same data*, the individual user experience may be different. The user interfaces are different, the manners of connection are different, and the application features are different. The Web version of the family calendar application runs on a server-class machine operated by a service provider. The service provider supports multiple underlying communication protocols. The Server calendar application implements most of the application logic discussed below. The family calendar also occasionally "consults" David's work calendar before scheduling personal appointments during work hours. David's company values work-life balance tremendously and has allowed limited access to his business calendar from his family calendar via a Web service interface.

Application Logic

The family calendar application has certain associated logic and semantics. The "rules" by which the application is guided include the following:

- Family events may include only some and not necessarily all members of the family.
- Events with overlapping times can be scheduled, provided that both events do not include the same family member.
- Scheduling conflicts are resolved by the Server calendar application using configurable *policies*, such as priorities of meeting types (e.g. ballet performances have priority over lunches) or priorities of meeting originator, (e.g. events scheduled by Mary have priority over ones scheduled by David).
- Certain events can be marked urgent. When an event is marked urgent, other family members are alerted to synchronize (in cases where they have Client devices that can be alerted), thus becoming aware of the urgent event.
- When family events are scheduled during normal work hours, the Server's calendar must request clearance from David's business calendar.

Clearly, a family calendar can have associated rules that are simpler or more complex than the above. The rules above are only illustrative.

Usage Instances

The following explores a few usage instances of this application and their possible realizations using different Client and Server application parts and SyncML as the underlying synchronization protocol.

Mary schedules a family dinner

Mary opens up the family calendar application on her PC. She enters a family dinner event in the calendar from 6:30 PM to 9:30 PM on the coming Friday. She then dials in to do some online shopping. The PC calendar application detects a connection and initiates a SyncML synchronization session with the Server calendar using the Hyper Text Transfer Protocol (HTTP). The Server calendar application accepts Mary's new entry, as it generates no conflict based on the rules above. Later, David, Susan, and Mark synchronize with the Server calendar and the calendars on their respective devices get updated with Mary's new entry.

Mark reschedules a soccer date

Mark just learned that his soccer game got moved from 3 PM Friday to 4 PM Thursday. David was scheduled to attend the game, taking off from work early on Friday. Mark updates the soccer game entry on his PDA application, marks it urgent for David to take notice, and synchronizes. The PDA application uses a SyncML Client to synchronize, using HTTP over a wireless dial-up connection. The Server calendar application receives the update and determines to check David's business calendar[1] and it checks the business calendar[1] and finds that David is available during the desired time. The Server calendar accepts Mark's change. Since Mark designated the update as urgent, the Server calendar uses the SyncML Server-alerted synchronization and alerts David to synchronize using the WAP Push [WPU01] feature on his mobile phone. David synchronizes upon the alert, becomes aware of the change, and is able to attend his son's soccer game.

Mary chooses ballet over lunch

Susan wants Mary to be present during her final rehearsal for an upcoming ballet performance. Susan therefore makes a new entry in the calendar application on her mobile phone, indicating the event from noon to 1:30 PM Wednesday. She then synchronizes with the

---

1. The business calendar is accessible as a Web service.

Server calendar. Her mobile phone calendar application uses SyncML with the underlying WAP transport protocol. The Server calendar accepts Susan's entry.

David and Mary sometimes get together for lunch during workdays. Unaware of Susan's new entry, Mary makes a new entry in the PC calendar application for lunch with David on Wednesday. When she synchronizes the PC calendar, the Server calendar application detects a conflict and, using a set of conflict resolution rules, resolves the conflict in favor of the ballet performance instead of the lunch appointment. The result of the conflict resolution is communicated in the same synchronization session. The PC calendar application processes the status message and indicates the ballet appointment in Mary's PC calendar. The application also chooses to communicate the conflict and its disposition to Mary, using a dialog box or other means.

The Benefits of SyncML

The use of SyncML to enable the above applications affords a number of benefits. We discuss the benefits from the perspective of the different parties involved in the realization and use of the application.

The consumer perspective

Imagine how you would feel if the hammer you have determines the nails you could use. Unfortunately, without SyncML, that is more or less an accurate characterization of mobile applications that require data synchronization. SyncML provides the user with the freedom to choose devices and service providers relatively independently of each other. Different members of the Stetson family use different devices according to their preference. The users are not compelled to use one particular device because a service provider only interoperates with that device. The user can also choose to change devices at some future time.

Consider a few examples of consumer flexibility. Mark may choose to use a mobile phone instead of a PDA if he feels the need for more spontaneous connectivity. SyncML also allows the Stetson family to choose service providers. In the future, if a different family calendar service provider offers a feature that they like, such as selective viewing of calendar entries to enable organization of surprise birthday parties, the Stetsons are free to switch to that service provider without having to buy a set of new devices. If another service provider offers a shared family picture gallery, the Stetsons are free to add that service provider as well.

SyncML enables the user to break free from artificial and cumbersome restrictions imposed by proprietary synchronization technology.

The device manufacturer perspective

For a company selling nails it is important that any hammer be able to drive those nails. If the nails can only be driven by one kind of hammer, that severely restricts the market for the nails. Similarly, for the manufacturer of David's mobile phone, it is important that the applications on the phone interoperate with the Server counterparts provided by various service providers. For the calendar application, the device manufacturer (or the application developer for a PC or PDA) can focus on user interaction and minimal required logic, such as keeping a record of local changes. The SyncML software on the device can handle the remaining mechanics of data synchronization by using standard data formats such as vCard [VCARD21], the SyncML Representation Protocol, the SyncML Synchronization Protocol, and HTTP or WSP [WSP01] transport protocols. Since the SyncML software on the device will interoperate with the corresponding SyncML software on the Server, the application on the device can work with applications on diverse Servers.

This is not necessarily true if there is a high level of *semantic* coupling between Client and Server application parts. In the family calendar application, most of the application logic is implemented on the Server. The Client part of the application is purposely generic and simple, enabling the Client application to work with diverse Servers. Having a common synchronization stack and simpler Client applications also reduces memory requirements on mobile devices.

The service provider perspective

For a company selling hammers it is important that their hammers be able to drive any nail. Similarly, it is important for service provider applications to interoperate with any device. By using the SyncML Synchronization Protocol underneath, the service provider's application can work with SyncML compliant applications on diverse Client devices. This enables broader market reach and penetration, driving revenue for the service provider.

The service provider can focus primarily on the application logic and semantics of data rather than trying to offer multiple synchronization protocols to suit the needs of many Clients. Using the specified SyncML transport bindings, the service provider can support Clients accessing the application via multiple transport protocols. The Stetsons use HTTP and WSP protocols from different devices. The service provider is also able to exploit certain characteristics of mobile devices and associated transport protocols by using additional features of SyncML, such as Server-alerted synchronization. The Server-alerted synchronization capability allows David to be quickly aware of the rescheduling of Mark's soccer game.

SyncML affords many advantages to the service provider. A family calendar application that correlates with the business calendar of a family member is much more valuable than one that cannot. Such functionality is orthogonal to SyncML and can be freely implemented by service providers.

Supporting Roving Nightingales

In the era of rising health-care costs, the City General Hospital provides extended care for its patients in their homes. The hospital employs a fleet of visiting nurses called the "Roving Nightingales." The nurses provide ongoing home care for some patients, as well as pre- and post-operative care for other patients. Over the course of a day, nurses visit a number of patients, record vital signs and other measurements, perform home tests, and collect samples for laboratory tests. The nurses also record a summary of their visual observation of the patients. Sometimes, if a patient is not doing too well, a nurse may decide that the patient needs to come in to the hospital-affiliated clinic to see a doctor. The hospital hosts a Complete eCare application that manages information flow and synchronization between the roving nightingales, doctors, and laboratories, as shown in Figure 3–3. In a real-life scenario, other entities like health insurance agents and pharmacies may need to be incorporated in the Complete eCare application. Those entities are ignored in this application for simplicity.

Application Setting

The Complete eCare application is the nerve center for the information flow and synchronization required for this scenario. Like the Web calendar, the various Client parts of this application run on various devices, such as a PDA, a laptop, an office PC, and a special purpose medical tablet device. Unlike the Web calendar application, however, the programmers that implement the Client parts used by the doctors and the

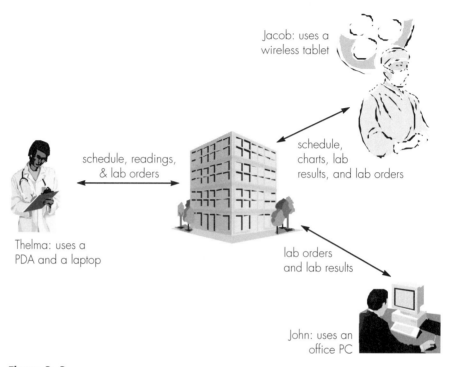

Figure 3–3
The setting for the Complete eCare application coordinating information exchange between different employees of the hospital and outside entities like independent laboratories.

visiting nurses also implement the Server part of the application. The information systems group of the hospital likely employs them. The Client part that runs in the laboratory office may be implemented by another independent organization.

Thelma, a typical roving nightingale, uses both a PDA and a laptop. She typically starts the day by synchronizing her daily schedule information with the Server calendar. She visits patients according to the schedule but synchronizes the schedule regularly to obtain schedule changes. When she visits a patient, she finds it convenient to record observations and home-test results quickly on her handheld PDA and synchronize the PDA with the Server's "daily visits" datastore. She finds the laptop more convenient for typing in visit summaries. At the end of the day, she usually synchronizes her daily visit records from the Server onto her laptop. The records already contain the information entered on the PDA throughout the day. She types in summaries for each visit and synchronizes the laptop to complete the records for the daily visits. Upon completion of the record, the laptop application also synchronizes the same record with the "patient chart" datastore. The patient chart datastore contains information in addition to records of the nurse visits. For certain patients that need to see a doctor, Thelma can mark certain records "red." When such red records are synchronized with the Server, the Server application schedules an appointment with a relevant doctor. Thelma can also collect samples (mailed separately to the lab) and order lab tests from her PDA. When the PDA is synchronized, the Server application updates the local "daily lab orders" datastore with the ordered test.

Jacob, a doctor in the hospital, uses a lightweight, touch-screen, multimedia-enabled, and wireless-LAN-capable tablet device. The device is his single-point portal for all the information that he needs to perform his job. He can access and make changes to his schedule, access and update patient charts, access lab results (including digitized x-ray images and CT scans), and order lab tests.

John, an office assistant in a small medical laboratory, uses a basic office PC. The PC is not normally connected to the network. Typically, the Client application on the PC dials in and synchronizes with the hospital Server in the morning to obtain the lab orders for the day. The orders are processed, lab tests are performed, and the results of completed lab tests are synchronized with the hospital Server at the end of the day. The data format of the lab orders and results conform to an XML-based standard called "LabML."[2]

The Server part of the Complete eCare application runs on server-class networked machines. The application manages many datastores, including schedules for nurses and doctors, lab orders and results, daily visit records for nurses, and patient charts. The Server supports the HTTP and WSP transport protocols over many underlying communication protocols including wide-area wireline and wireless, as well as local-area wireless.

Application Logic

Most of the application *intelligence* resides with the Server application. For illustrative purposes, the Server application may be guided by the following rules:

---

2. For the time being, LabML is imaginary. It is reasonable, however, to expect the emergence of domain-specific standard XML formats in the near future.

- The Server application may enter appointments, perhaps following a call into the office from a patient. If there are conflicts between appointments entered by the application and those entered by the doctor or the nurse, the appointments entered by the doctor or the nurse take precedence.
- If the nurse and the doctor have ordered the same lab test for a particular patient in the span of a week, the second test order is discarded, and the doctor or the nurse is notified by email.
- Ordinary entries from the visiting nurse are synchronized with the daily visit datastore. Entries marked "red" are also synchronized with the patient chart datastore. When an entry marked "red" is received, the Server application also schedules an appointment for the patient to visit the doctor at the hospital clinic and alerts the office assistant to notify the patient about the appointment.

Usage Instances

The following illustrates a few usage scenarios where SyncML is used to synchronize different data between a variety of devices using diverse network protocols.

Thelma visits her first patient

Thelma turns on her WAP-enabled PDA in the morning. She had manually entered a morning appointment on her schedule the previous evening. Upon synchronization, the Server application discovers a conflict with a previously scheduled appointment and resolves the conflict in favor of the appointment entered by Thelma. The conflict, however, is flagged during synchronization and the conflicting record is returned to the PDA. The PDA application brings the conflict to Thelma's attention and she enters a new appointment in the afternoon to visit the originally scheduled patient.

Thelma visits the first patient. She records vital signs and performs a few quick tests, and then enters the observations and results in the local PDA. The patient appeared fine and there was no need to order any lab tests or to schedule a doctor's visit. She synchronizes her PDA to quickly upload the visit records to the Server. She does this to ensure that the records are not lost in case the PDA is lost, stolen, or broken. At the end of the day, Thelma synchronizes her laptop application with the Server daily visit datastore. Her laptop application obtains the daily updates previously synchronized from the PDA. She fills the descriptive visit summary in the empty summary field and resynchronizes the laptop to complete this visit record and other similar records.

Thelma's lab order application is upgraded

The "LabML" standards committee had introduced a new field called "priority" in the lab order data format. The priority field is intended for use by the requester to express priorities of high, medium, or low. A laboratory can now better schedule its lab tests according to the expressed priorities of the orders. Over the last few weeks, the Complete eCare application programmers have updated the Client and Server applications to expose this new field in the Client GUI and make necessary changes in the data formats. Now the updated applications are ready for deployment.

The SyncML Device Management Protocol (see Chapter 9) is ideal for configuring and managing software on diverse client devices. When Thelma synchronizes her PDA after visiting the first patient, the Client information technology (IT) management application on Thelma's PDA uses the Device Management Protocol to obtain the latest updates to the Lab Order application code from the hospital Server. It installs the application code and enables Thelma to take advantage of this new feature immediately. When Thelma connects her laptop to the Server in the evening, the Client IT management application on the laptop similarly obtains the latest updates to the Lab Order application code from the Server.

Thelma's second patient is not doing too well

Thelma visits her second patient of the day and finds that he is not doing too well. The patient is under post-operative in-home care and his recovery is not progressing as expected. She records all the measurements and marks the daily visit update "red"; she also orders a high priority blood test on her PDA application, and then synchronizes her PDA. The Client application synchronizes the daily visit update with not only the Server daily visit datastore, but also the Server patient chart datastore, since the entry is marked "red," The Server application schedules a next-day afternoon appointment with the appropriate clinic doctor. The Server application also synchronizes the lab order with the Server lab order datastore. Meanwhile, just before lunch, Thelma ships the blood sample in a protective envelope for next-day morning delivery to the lab.

The doctor, in the middle of his morning rounds, synchronizes his wireless tablet calendar with the Server calendar. He notices the new appointment (during his clinic office hours) with the patient. He opens his local patient chart datastore and synchronizes it with the Server. The synchronization updates the local patient datastore with the latest patient chart from the Server datastore for this patient. The doctor reviews the chart briefly over lunch the next day in preparation for the afternoon visit.

The lab results are ready in time

In order to differentiate from the competition, the laboratory has instituted a policy that results for high priority lab tests will be synchronized with the hospital every two hours. John synchronizes the laboratory office PC with the hospital lab order datastore at 5 PM and picks up the high priority lab order that Thelma entered. The corresponding blood sample arrives the next morning. The sample, being high priority, is tested quickly. The results are ready by 1 PM. John synchronizes the results for this high-priority order in the 2 PM batch and the results appear in the Server lab results datastore. The doctor had noticed the outstanding lab order during his lunch review of the patient charts. Just before the patient visit at 3 PM, he synchronizes the lab results datastore on his tablet with the Server datastore and obtains the results. He is now fully informed before the patient visit and is able to make the best medical decision for the patient.

The Benefits of SyncML

The use of SyncML in enterprise applications is beneficial from the point of view of the employees, the platform vendors, and the enterprise IT personnel, such as application programmers and system managers.

The employee perspective

Some of the biggest benefits from the employee standpoint are flexibility, empowerment, and usability. Since enterprise applications are usually richer in function and semantics, it is conceivable that certain parts of the application are best performed on certain devices. In Thelma's case, she finds it convenient to record brief results of observations and home-tests in her PDA as she is performing them. This way she does not have to take them down on paper and later re-enter them. On the other hand, entering lines of text for the visit summary is usually difficult to do on a PDA-type device. Thelma finds it convenient to record her subjective summary later in the day on her laptop. SyncML allows synchronization from diverse devices, allowing the flexibility that Thelma desires. SyncML allows applications to detect and resolve conflicts but enables communication of conflicts in a standard way in its Synchronization Protocol. Policies such as "Client resolves the conflict" can be implemented with relative ease by applications. Thelma is notified and allowed to resolve scheduling conflicts herself directly from her PDA, thereby feeling more empowered as an employee to manage her daily schedule. Thelma is also not worried about managing applications on her mobile devices. SyncML Device Management is used to seamlessly upgrade applications on her device in a timely fashion. This increases overall usability from her perspective and makes her a more effective roving nightingale.

The platform vendor perspective

In the previous Web calendar example, independent individuals make the buying decisions for the server platform and the client platform. In contrast, in the enterprise case the client and server platforms are bought by the same entity, namely the enterprise management, perhaps advised by the enterprise IT group. Without synchronization standards, it is possible that a dominant server vendor will also drag in client platforms that it owns, or its partners own, by closely coupling the platforms together technically. The reverse is also true. From a technical perspective, a synchronization standard such as SyncML allows server and client platform buying decisions to be relatively independent of each other. For example, when an enterprise buys a server platform, maybe the very best server platform, the owners of that platform cannot drag in its clients by claiming that one must buy these clients to communicate with this wonderful server. The world has converged largely on the TCP/IP communication protocol, and therefore any client implementing the TCP/IP protocol should be able to communicate with any server that does. This allows the clients to freely compete on the client side of the business. SyncML takes this argument to the next level, from the communication layer to the data exchange layer. By adopting a standard data synchronization protocol, the clients and servers can freely compete on the merits of their individual implementations and not use one leverage point to establish another.

The enterprise IT perspective

SyncML is enormously beneficial from the enterprise IT perspective. First, it reduces the cost of developing mobile enterprise applications. Both Client and Server application programmers can depend on the underlying SyncML stack. Assuming some form of a common SyncML toolkit that allows generation, processing, and communication of SyncML packages over a variety of transports, the application programmers can freely focus on the application logic and user interface issues. The SyncML toolkit can either be developed by the enterprise IT team or bought from a tool vendor, or can actually be included with the platform, as SyncML becomes more commonplace. Once there, the toolkit can be used in common by various applications. Therefore, although it is likely that enterprise applications will use proprietary data formats between Client and Server parts and will likely not be targeted to interoperate with applications developed by other entities, the use of SyncML as an underlying synchronization mechanism will reduce overall application development time.

Enterprise applications also need to synchronize with other enterprise applications. For example, the Complete eCare application includes synchronization with a Client in an independent laboratory. In order to really interoperate with applications developed across enterprises, standard data formats and their semantics need to be defined and used by SyncML. The imaginary LabML standard data format is key to interoperability between the hospital and the laboratory. Using a standard synchronization protocol, such as SyncML, and a standard data format, such as LabML, the hospital is able to interoperate with various independent laboratories. Thus, using SyncML, applications within an enterprise are well positioned to interoperate with applications outside the enterprise as standard data formats emerge.

Managing client devices is a key element of the overall IT cost in many enterprises. The software components of various devices, including application software, change from time to time. It is difficult to "roll out" the new versions of the software components in a consistent manner. This problem, however, is similar to data synchronization, where the "data" is the binary code for the software component. The IT management applications can use the SyncML Device Management Protocol (which uses the Representation Protocol) to keep device software and applications consistent. In the above example, Thelma's PDA gets updated with the new application code in a timely fashion, enabling a chain of events that result in better overall patient care. Chapter 9 covers SyncML Device Management in more detail.

The Reach of SyncML Applications

SyncML is aimed to enable truly interoperable data synchronization. In a world where applications can synchronize and exchange data with each other in an interoperable fashion, a floodgate of applications is suddenly opened. The rows of Table 3–1 list various vertical industries. The columns list various cross-industry functions. SyncML potentially enables applications in *every individual cell* of this table. This chapter has already offered examples above for Internet services, and health care services and operations. We list a few more examples below and check off the corresponding cells.

- An auto insurance adjuster moves around town inspecting several vehicles for collision damages and writing adjustment reports in his laptop. Later in the day, the reports are synchronized with the Server datastore. The adjuster report is later correlated with filed claims to determine the amount of reimbursement (Insurance/Operations).
- A sales representative for a pharmaceutical company goes around to various drugstores in a city. He secures orders for drugs and other pharmaceutical supplies over the course of a day and records them in his handheld PDA. At the end of the day, he synchronizes his PDA with the Enterprise Server and the processing of the orders begins (Health Care/Sales & Distribution).
- An employee in a local department store inspects the aisles and makes entries in his handheld PDA for store items that need replenishment. He then goes in the back office, and synchronizes his PDA with the store server (a PC). The store server places restocking orders for the appropriate items (Retail/Procurement).

As an exercise, the reader is invited to reflect and imagine a few more applications and populate a few more cells of the table. The reader will likely succeed in populating the table substantially in a relatively short time. The applications in the table are in fact applications of data synchronization and not strictly applications of SyncML. The key point to note is that by enabling interoperable data synchronization, SyncML enables a wide variety of business applications.

Table 3–1

The rows identify various vertical industries. The columns identify cross-industry functions. The checkmarks indicate applications above.

	Sales & Distribution	Marketing	Supply Chain	Manufacturing	Customer Relations Management	Procurement	Services	Operations
Health Care	✓						✓	✓
Online Services							✓	
Automotive								
Electronics								
Energy								
Chemical								
Banking								
Insurance								✓
Retail						✓		
Travel								

52

Part II

SYNCML IN-DEPTH

4

SyncML Fundamentals

SyncML® is an ambitious endeavor. It attempts to formulate a data synchronization protocol that interoperates across devices, networks, data types, and applications. Although the development of a common interoperable de facto synchronization standard is its primary goal, it must be designed such that it can be efficiently realized. Among other things, efficiency includes coping with resource-constrained mobile devices, operating over low-bandwidth wireless networks, and enabling servers to support millions of mobile devices. A data synchronization standard that interoperates but does not perform well adversely affects the overall user experience. Such a standard is not likely to be widely adopted. These considerations form the design goals of SyncML. The first of the following sections outlines the SyncML design space and discusses specific design goals of SyncML, along with the approaches adopted by SyncML to facilitate these goals. The second section provides an insight into the SyncML architecture.

The Design Goals of SyncML

The design space of SyncML is large but not intractable. Figure 4–1 outlines the design space of SyncML. In reality, it has more dimensions than just the four shown here. The dimensions shown are deemed the most important, as they have considerable influence on the design decisions. They are device, network, data, and synchronization topology. The device dimension is in the increasing order of resource richness and capabilities. On the lower end, there are devices such as cellular

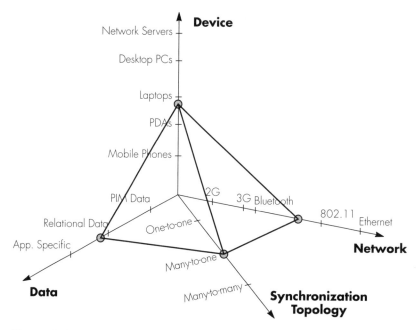

Figure 4–1
The primary design dimensions of SyncML and its overall design space. The space enclosed in the pyramid is the key design thrust of SyncML.

phones and PDAs; on the higher end there are personal computers and server-class devices. The network dimension is in the increasing order of bandwidth and decreasing order of latency. On the lower end we have wide-area wireless networks, such as cellular networks, and local low-power wireless networks, such as Bluetooth™ [MB01].[1] On the higher end we have wireless local-area networks, such as IEEE 802.11 [WLAN02] and regular wireline networks, such as Ethernet. The data dimension includes widely adopted PIM data on the lower end, relational data in the middle, and application-specific data on the higher end. As outlined in Chapter 1, data synchronization may occur among various entities conforming to certain synchronization topologies. The topologies supported have fundamental implications on the design of a synchronization protocol. The topology dimension includes one-to-one synchronization on the lower end and many-to-many synchronization on the higher end.

The pyramid in Figure 4–1 identifies the design thrust of SyncML. In summary, SyncML is optimized for mobile devices such as mobile

---

1. Some 3G networks can have higher bandwidth than Bluetooth.

phones and PDAs; SyncML accommodates the characteristics of low-bandwidth, low-reliability, and high-latency wireless networks; SyncML attempts to efficiently support common data types, including PIM data and relational data; and SyncML efficiently supports one-to-one and many-to-one synchronization topologies. It is important to note that the design thrust only indicates what SyncML is intended to support best. SyncML neither rules out nor handicaps operations outside its design thrust. For example, one can gainfully use SyncML to synchronize two servers, but the specification does not take any special steps to make such synchronization efficient. Similarly, one can use SyncML to perform synchronization over a local area network (LAN) or perform synchronization of non-standard or application-specific data. Below we discuss a few specific design goals of SyncML.

Effectiveness over Wireless Networks

Numerous target SyncML devices, such as mobile phones and PDAs, are usually connected wirelessly. Mobile phones are connected via various 2G and 3G wide-area wireless networks. Some PDAs also connect using these networks. In addition, some PDAs and mobile phones will begin to connect using local area wireless technology such as Bluetooth™ and IEEE 802.11 as these technologies become more prevalent. It is therefore immensely important that SyncML be designed to be effective over wireless networks. Although there is a large set of wireless networks used by mobile devices, the following characteristics of wireless networks are generally true when compared with regular wireline networks, such as Ethernet or T1-based wide area networks:

- Limited network bandwidth
- High network latency
- Low reliability
- High communication costs

SyncML attempts to address all the above issues in the design of its components. The following outlines some pertinent design approaches.

Judicious use of bandwidth

XML [HM01] can be verbose. The Wireless Application Protocol (WAP) [ACS+00] forum has defined a standard for binary representation of XML documents called WBXML [WBXML01]. SyncML allows the use of WBXML to transmit messages. Binary encoding reduces overall transmission requirements.

At the cost of complexities in mobile Clients, SyncML encourages Clients to maintain some kind of change log to account for changes made to local datastores. Thus in general, synchronization between a Client and a Server is incremental, involving changes since the last synchronization. This reduces bandwidth requirements tremendously compared to complete synchronization of all datastore entries in the Client. Such complete synchronization is used in SyncML only to handle exceptions such as first-time synchronization or resynchronization after a failure.[2] Although common SyncML applications may exchange complete data items, SyncML allows the communication of changes to data items instead of complete data items, thereby preserving bandwidth. One can envision the emergence of standard data formats, such as for Relational Data, allowing specification of only the changed columns in a database row. As such formats emerge, SyncML can leverage them to reduce the overall bandwidth requirements of synchronization.

Combating network latency

Wireless networks usually have high network latency. In an environment with high latency, one must avoid a chatty protocol at all costs. In a chatty protocol, individual data items and/or operations are typically communicated separately. In contrast, SyncML allows the batching of data items and operations in one message. Batch transmission of data and operations masks network latency to a large extent, as the processing of a batch of items can continue while the next batch is in transmission.

Addressing low reliability

While batching is beneficial in combating latency, a large batch or Message may not make it across a wireless network in its entirety. This is because wireless networks are relatively unreliable, and it is quite likely that errors or failures may occur during the transmission of a large package. Although this problem may apparently be deemed a low-level network layer problem that can be addressed by packet protocols, TCP/IP (and hence HTTP), one of the common "reliable" protocols that applications use, is unfortunately not especially suited for wireless networks, as it does not allow for incremental progress of transmission. TCP/IP may transfer 699 bytes of a 700-byte message 100 times and still declare overall failure of message delivery. It lacks the ability to "pick up the transmission where it left off" (incremental

---

2. This is called "slow synchronization." See Chapter 5.

progress) and just send the remaining one byte in the second round after sending 699 bytes successfully in the first round.

Another common transport, Wireless Session Protocol (WSP) [WSP01], uses WAP gateways that can limit the maximum size of messages. A WAP gateway will not allow a large SyncML Message to pass through. In light of these factors, SyncML allows partitioning of a logical package into smaller physical messages. In such a situation, SyncML implementations become more complex, as they now need to support package assembly and disassembly. Overall, it is a good tradeoff, however, as multiple messages allow SyncML to combat lower reliability and other configuration limitations (e.g., WAP) of wireless networks.

Reducing communication costs

The cost of communication is commonly assessed by the amount of time and/or the amount of data communicated. The combination of WBXML compression, to reduce the number of bytes communicated, and batching, to reduce the overall latency of communication, is aimed at substantially reducing the overall communication costs.

Support Transport Heterogeneity

Not only are there various kinds of physical networks, such as wired and wireless, that differ in characteristics, there are also various kinds of transport protocols that are used over these networks: HTTP, WSP, and Simple Mail Transfer Protocol (SMTP) to name just a few. The SyncML Synchronization Protocol is at a higher semantic level than any of these transport protocols. It is therefore imperative that SyncML be realizable over many transport protocols. From the SyncML perspective, which is that of an "application" on top of the various transport protocols, the protocols offer the following different classes of behaviors:

- Persistent connection
- Synchronous request-response
- Asynchronous request-response

Examples of protocols that offer a persistent connection are TCP/IP and OBEX [OBEX99]. TCP/IP is one of the most commonly used protocols. The OBEX protocol is used for data exchange by the infrared (IrDA®) and Bluetooth communication technologies. In these protocols, the two communicating parties actually maintain a long-lived connection over which all communication occurs during synchronization. Therefore, the communication has *state*. This implies that the communicating parties can exchange multiple messages pertaining to a synchronization session efficiently over the same connection. All the associated context of the synchronization, such as authentication, is implicitly attached to the existing connection. In addition, these protocols are symmetric such that any communicating party can initiate synchronization.

HTTP and WSP are popular synchronous request-response protocols. Although HTTP is implemented on TCP/IP, it offers a request-response abstraction to applications. In these protocols, a "client" sends a "request" to a "server" and the server returns a "reply" synchronously. This request-response paradigm works well for a model where clients and servers are substantially asymmetric and clients generally consume information that servers generate. In this model, usually the client initiates synchronization. Specific protocols like WSP, however, support "push," whereby a server can send an alert to a client, indicating that the client should begin synchronization.

In addition to the above, there are asynchronous request-response protocols. Although these protocols are also fundamentally request-response in nature, the response is not sent as an in-band reply to a request. The response is sent out-of-band after an undetermined time period. Popular email protocols such as SMTP, POP3 (Post Office Protocol), and IMAP (Internet Message Access Protocol) are of this nature.

SyncML is intended to run over all the above diverse protocols. The Synchronization Protocol is designed such that it is simply a specification for a series of packages (realized as physical messages) exchanged between a Client and a Server in some order. As discussed above, SyncML allows a package to be communicated as several transport-layer messages. SyncML, however, does not depend on the underlying transport protocol to support any ordering of these messages but enforces ordering itself by sequencing messages. The sending party must receive implicit or explicit acknowledgment for one message of a multiple-message package before it is allowed to send the next message for the same package. The message exchange sequence of SyncML can be easily and efficiently mapped to common synchronous request-response protocols such as HTTP or WSP.

The Synchronization Protocol can also be mapped to asynchronous request-response protocols such as SMTP; however, an implementation will require the client and the server to do frequent polling (analogous to repeatedly checking email) and may not be so efficient.[3]

---

3. To avoid polling, notification of message arrival can be achieved by using orthogonal means, such as ISDN D-channel. This, however, requires additional capabilities and introduces more complexity.

Therefore, although SyncML supports SMTP-like protocols, that is not within the recommended "operating range" of the design.

Protocols such as TCP/IP and OBEX can support SyncML message sequencing in a straightforward manner. Actually, over a persistent connection, message sequencing is not necessary, as the protocol itself will preserve ordering over the same connection. Requiring SyncML message sequencing over these protocols is slightly suboptimal design. SyncML designers knowingly made this tradeoff, as most emerging popular Internet protocols are of the synchronous request-response type—and message sequencing is a necessity if SyncML is to operate over such protocols.

The mechanics of using various transport protocols in SyncML is straightforward, as discussed in Chapter 7. For each transport protocol, SyncML requires the specification of a *binding* that determines how a SyncML package is realized into one or more physical messages over that transport. SyncML defines bindings for popular protocols, such as HTTP, WSP, and OBEX. Other bindings can be defined and used as deemed appropriate.

Support a Rich Set of Networked Data

Since SyncML is aimed at enabling data synchronization for a diverse set of mobile applications, it is natural that it be able to support a diverse set of networked data. In particular, SyncML should support synchronization of the following kinds of data related to popular applications:

- Data related to personal information applications, such as contact, calendar, and to-do list information
- Data related to collaborative applications, such as email and network news
- Data related to Relational Database applications
- XML and HTML documents
- Any other data represented as a MIME (Multipurpose Internet Mail Extensions) [RFC2045] type

Consider a scenario where an application vendor provides the Server-side application, the Client-side application, and synchronization between these applications. An example of such a scenario is the set of Palm® personal information management applications. In such situations, the format of the Client-side data, the Server-side data, and that of the data exchanged during synchronization can be entirely private. The application vendor can completely control the data format. The data format can be optimized for storage on the Client and the Server, as well as for transport over the communication link. Now consider a scenario where Client and Server applications are synchronized using software provided by an independent synchronization vendor. The Client and the Server application vendors (sometimes the same entity) have to provide their data format information to the synchronization vendor. The synchronization vendor still reserves the right to transmit the data in a proprietary format during synchronization, as it controls both ends of the synchronization. This approach does not work in SyncML. For truly interoperable synchronization, it is imperative that the data exchange format between the Server and the Client be open and standards-based. The entity that writes the Client synchronization agent may be different than the entity that writes the Server synchronization agent. The only way they can interoperably exchange data is by using standardized formats.

Unfortunately, open standards for data formats, or at least widely adopted formats, do not exist for all the data that may be potentially synchronized using SyncML. Standards for common data formats, such as contact, calendar, and to-do list information, exist and SyncML simply adopts those standard formats. An interoperable SyncML implementation must use the prescribed data formats. For example, the vCalendar [VCAL] or iCalendar [RFC2445] data format should be used for the calendar applications. Standards for Relational Data do not exist but are emerging. Standards for other XML data in general do not exist but are also emerging. The growing interest in XML Schemas will likely act as a catalyst in the standardization of various kinds of data. SyncML adopts an approach whereby data standards are incorporated in SyncML as they emerge.

While SyncML recommends that standard data formats be used for certain applications and enforces the statute for SyncML Servers in the interoperability testing process (Chapter 13), it does not preclude the use of proprietary data formats by applications. It is important for applications to have the choice of using proprietary data formats. Some applications, for example, may like to synchronize arbitrary binary data. Such applications should not be prevented from using SyncML. Implementations that use nonstandard data formats, however, are not certified as interoperable.

It is important to note that the requirement of using standard data formats for synchronization does not mean that the Client and Server applications are forced to use the standard data formats internally within the applications. The calendar application in a mobile phone likely stores calendar data in an efficient binary format. A Server calendar application can also choose to store data in its own efficient format. During synchronization, however, the respective synchronization agents must transform these formats into the standard format and vice versa.

Neutrality to Programming Environments

A de facto standard for mobile data synchronization has to operate over a variety of programming environments. A programming environment consists of the programming language, implied system resources, such as files or datastores, and processing capabilities, such as single or multi-threaded. The programming environment also consists of the networking environment, which is covered above and hence ignored here. In a mobile phone, the available programming language could be the C language, there may be no file system or datastore abstractions but proprietary native ways of storing data, and there may be no facilities for multithreading. In a PDA, the available language could be C++, there may be a file-system available, and multithreading may be supported (e.g., PocketPC-based PDAs). In a network server, languages such as C, C++, and Java™ may be all be available, both file system and datastore abstractions may be available for storage, and there may be rich support for multithreading.

SyncML does not make any assumptions about the programming language supported by a particular platform. It is based on exchanging well-specified, structured XML messages and not on any particular programming environment. The specification only determines the format of the information that is exchanged and the sequence of information exchange. The information exchanged (SyncML packages) can be generated in any way deemed appropriate by a programmer. The reference implementation that accompanies the SyncML specifications is based upon a published C API, but the API per se is not part of the standard.

The issue of neutrality is deeper than just the issue of programming language. By virtue of simply being XML-based, a certain amount of neutrality is achieved. For example, in a network Server, a SyncML message can be processed using a Document Object Model (DOM) [DOM02] tree in parallel with multiple threads traversing or constructing different parts of the tree. In a mobile client where no multithreading support is available, the same message can be serially processed using the Simple API for XML (SAX) [SAX02], which does not require efficient processing of a parse tree.

SyncML is also neutral to available platform storage abstractions. Object-based storage abstractions such as Object Databases are actually well suited to store, retrieve, and directly manipulate SyncML (XML) documents. Such storage abstractions allow the programmer to directly store the in-memory representation of an XML document. Neither common Relational Databases nor file systems offer object-based abstractions for storing XML documents. In most cases, however, the SyncML document is transiently generated and processed in memory during synchronization in such a way that the document itself is not stored persistently. The data within the document, such as a calendar entry, is of course processed and then stored using the optimized internal representation suited to a particular platform.

Support Multiple Synchronization Topologies

There are three kinds of synchronization topologies[4]: one-to-one, many-to-one, and many-to-many. In the one-to-one data synchronization model, a particular client only synchronizes with a particular server. In the many-to-one data synchronization model, two or more clients synchronize with a single server. In the many-to-many data synchronization model, a group of computers freely synchronize with each other directly. In many-to-many synchronization, there is no notion of a primary server or datastore.

The one-to-one interaction model and the many-to-one interaction model are abundantly more common in the context of day-to-day commercial applications. Moreover, the many-to-many model can be indirectly (but inefficiently) achieved using the many-to-one model. This is done by designating one device as a Server and stipulating that the other devices synchronize with the Server and hence indirectly synchronize with each other via the Server. Implementing the many-to-one interaction model (which includes the one-to-one model) is conceptually simpler, and the resulting implementations are orders of magnitude simpler than ones that support the many-to-many interaction model. In many-to-many synchronization, complex data structures such as "version vectors" need to be associated with data items to correctly synchronize data. Maintaining consistency of data identifiers in many-to-many

---

4. See Chapter 1 for a more complete discussion of synchronization topologies with usage examples.

synchronization is also complex without forcing all parties to store identifiers in the same format. The many-to-many model is also especially intractable for the purposes of accounting and failure recovery.

For the above reasons, SyncML is optimized for the many-to-one synchronization topology. It allows the exchange of datastore sync anchors[5] in the beginning of a synchronization, which indicates the last "timestamp" at which the two computers synchronized. The timestamp could be an actual time value or a logical counter. Based on the exchanged sync anchor values, the associated sync engines could use simple data structures such as change logs (see section below) to determine which data items have changed since the last instance of synchronization.

SyncML, however, allows many-to-many synchronization. It allows each data item to have an associated version, which could actually be a version vector required for many-to-many synchronization. It also does not specify the format of the sync anchor explicitly, so the sync anchor could also be a version vector. Furthermore, it allows a SyncML device to play the dual roles of a Server and a Client. The above allowances imply that SyncML can be used for most advanced forms of data synchronization, such as many-to-many synchronization, but is optimized for the more common case of many-to-one synchronization.

Address the Resource Limitations of a Mobile Device

A typical mobile device, such as a cellular phone or a PDA, is resource-limited in many ways. The key limitations include the following constraints:

- Limited memory
- Limited processing capabilities
- Limited battery power
- Limited communication capabilities

The constraints on communication arise from the fact that these devices often use wireless networks. The design considerations associated with wireless networks are discussed above and thus are ignored in this section.

---

5. See Chapter 5 for a more comprehensive discussion of sync anchors and their usage.

Addressing memory limitations

The SyncML specification is extremely sensitive to the implied memory requirements. The *static footprint*, or overall code size, of the SyncML implementation on a mobile Client device should be low—on the order of tens of kilobytes. SyncML does not require devices to validate received packages against the SyncML DTD. Checking syntactic correctness is deemed enough. This allows devices to use simple parsers instead of the more complex parses used to validate XML, which are usually of much larger code size. Although SyncML allows a device to play the dual roles of Client and Server, it normally expects that devices will play only single roles. SyncML assumes that mobile devices will most often play the Client role. In the Client role, many features are made optional that would have to be supported in a Server role. By consciously allowing devices to announce their roles and by having asymmetric requirements for those roles, SyncML allows for focused, lean implementations on mobile Clients.

The *dynamic footprint* is the overall memory required by a program when the program executes. SyncML is designed to reduce the overall dynamic footprint required during the process of synchronization. It allows Client data identifiers to be smaller than those in a Server. It consciously burdens Server implementations with the task of identifier mapping (see Chapter 5) between various Client identifier formats and the Server identifier format. This enables a Client to use very compact, optimized data identifiers, thereby reducing dynamic memory requirements. Clients can also use SAX parsers for parsing SyncML documents instead of DOM parsers, as SAX parsers require considerably less execution-time memory than DOM parsers. The amount of communication buffer space required during synchronization is a key aspect of dynamic footprint requirements. By allowing packages to be broken into multiple smaller messages, SyncML reduces the communication buffer space required at any one time. SyncML also allows a Client to specify the largest message size that it can process and thereby allows the Client to adapt to its current available memory.

The memory overhead associated with data synchronization includes maintaining a *change log*. The change log is a logical name for information that a synchronization engine or an application must maintain corresponding to each datastore. The role of the change log is to record which items have changed in a datastore between successive synchronizations. During synchronization, it is expected that the change log be consulted to determine what pertinent changes must be communicated. The actual implementation of a change log may take various forms. It can actually be a physical *log* of operations that have been made to items in a datastore. Each log entry may contain the type of operation and a timestamp. The timestamp could be an actual time or some logical counter maintained by the synchronization framework. In cases where a datastore in a client device is associated only with one datastore on one server, the change log can also take the form of a *change bit* associated with every data item. The change bit approach for mobile devices may conserve device storage in some cases where the device is synchronized relatively infrequently and the log-based change log can potentially grow substantially between synchronizations. However, for a client datastore that is synchronized with multiple datastores, a simple change bit does not suffice. Change bits have to be maintained for every synchronization partner for every data item. The storage requirements explode and become infeasible for many mobile devices. For many common synchronization usages, it often makes sense to use an actual log-based change log. In a log-based approach, however, the change log grows continuously and must be pruned from time to time by deleting entries corresponding to changes that have been communicated to all synchronizing partners.

Addressing limitations of processing capabilities

SyncML is sensitive to the limitations of the processing capabilities of typical mobile devices. SyncML encourages the Client to be simple and encourages that detection and resolution of conflicts and interpretation of application data be done on the Server. Also, the ability to opt for a simpler, nonvalidating parser is of substantial assistance to such devices. Activities related to security also tend to be processor-intensive. SyncML allows Clients to use simple password authentication schemes, which do not require much Client-side processing. Higher-end Clients can still use encrypted message digests for authentication. In its current version, SyncML does not mandate any data encryption requirements,[6] as data encryption and decryption tends to consume substantial processing power.

Addressing battery power limitations

Battery power is a precious resource for many mobile devices. By addressing processing capability restrictions, SyncML partially addresses

---

6. Data encryption, although not mandated, is allowed. For example, with the HTTP binding, it is possible to use SSL (Secure Sockets Layer) or TLS (Transport Layer Security).

battery power restrictions as well. Another battery-saving feature of SyncML is server-alerted synchronization. Common usage examples of synchronization often involve mobile clients obtaining data updates from a server. For example, field insurance agents may want to obtain the latest rate quotes, or mobile health-care professionals may want to obtain the latest test results on a patient that they are about to visit. If the client always initiates synchronization, it has to poll the server at certain intervals to get the latest server updates, not knowing if there are any pertinent updates. This wastes battery power (along with the user's time and money). Therefore, for certain clients, such as mobile phones, SyncML allows a Server to *alert* a Client to begin synchronization. Such an alert function may be implemented on the underlying "push" functionality of a transport protocol, when available (e.g., WSP). Allowing server-alerted synchronization potentially conserves the battery power of a mobile device, in contrast to repeated polling. SyncML also does not require that the device or a particular datastore be "locked-out" during synchronization. It allows reads and updates to continue as synchronization progresses. By allowing the user to perform productive work while synchronization continues, it optimizes overall usage of battery power, as well as preserving the continuity of user experience.

Allow Building of Scalable Servers

Typically, SyncML Servers are expected to serve a large number of Clients. For an Internet service provider, the SyncML Server is expected to serve hundreds of thousands of Clients, of which tens of thousands may be simultaneous users. For an enterprise Server, the SyncML Server must serve thousands of Clients. Therefore, it is of critical importance that SyncML Servers be scalable. The SyncML specification takes a few explicit steps to allow for the scaling of SyncML Servers.

Batch processing of data

As indicated above in the discussion regarding latency of wireless communication, SyncML encourages "batch" processing as much as possible. In batch synchronization, the Client typically sends changes made to one or more datastores in a single SyncML package. The processing of data in this manner also allows the building of scalable Servers. First, during the process of synchronization, the Server typically accesses some back-end datastore, such as a Relational Database or a PIM datastore (e.g., Lotus Notes®, Microsoft Exchange®). Accessing these datastores and performing multiple operations simultaneously usually provides much better performance than individual unit operations. Second, a SyncML Server may not only batch updates from one synchronization session with one Client, but may do so over multiple sessions with multiple Clients at the same time. This kind of collectively batched operation could dramatically increase throughput with the back-end datastore. Third, batching amortizes constant processing costs associated with SyncML Messages. For example, authentication could be performed just once for one SyncML Message that contains operations for one user across multiple related datastores (assuming the user uses the same credentials for those related datastores).

No implicit ordering constraints

SyncML does not require that commands within a SyncML Message be processed in any particular order.[7] Commands pertaining to the same or different datastores can be processed in any order. This enables a Server to process a SyncML Message using multiple concurrent threads without appreciable timing coordination among the different threads processing a SyncML Message. A network Server can therefore have hundreds of threads processing SyncML Messages concurrently, thereby increasing overall performance and scalability.

No transactional guarantees

SyncML does not guarantee any transactional semantics[8] with data synchronization. For example, consider an `Add` operation from a Client. When the Server synchronization agent processes the `Add` operation, it can wait until the operation is confirmed by the back-end datastore or it can simply "queue" the operation to another entity that manages back-end operations and move on to the next operation. In the first case, the synchronization agent is guaranteed that the operation is actually complete at the back end (for back ends that provide transactional guarantees, e.g., a Relational Database). In the second case, the synchronization agent does not actually know if the operation has been completed. The first case is clearly more time consuming and could delay synchronization, adding to overall latency in processing Messages. The second case enables quicker processing of SyncML Messages but incurs the risk of back-end rejections or failures after the synchronization is deemed complete. Clearly, back-end failures can occur for multiple reasons. Back-end datastores can also reject updates for numerous reasons, including

---

7. One exception is commands that are enclosed within the `Sequence` command. See Chapter 6.
8. One exception is the `Atomic` command. See Chapter 6.

specific constraint violations such as adding an employee record that indicates a salary of one billion dollars per month. Since SyncML does not provide transactional guarantees, it enables Server implementations to adopt the second approach. In such an approach, after-the-fact back-end failures could simply be treated as new back-end updates during the next round of synchronization. SyncML expects such after-the-fact failures to be relatively few and makes a conscious tradeoff in favor of concurrency, performance, and scalability.

Enable load balancing

Load balancing is a common technique employed in building scalable systems. In this technique, a group of physical server machines act as one logical server. Typically, when a request from a client comes into the system, it is routed to an appropriate, lightly loaded physical server. The "router" balances load across multiple physical servers, thereby increasing the scalability of the overall system. For load balancing to work best, it is important that there not be much history in between two requests from a client. If two requests are strongly correlated and access common intermediate "state" on the server, those requests should be routed to the same physical server. In SyncML, the coordination between two synchronization sessions with one Client is encapsulated in the sync anchors pertaining to each datastore. Sync anchors are concise and hence can be shared between physical Servers using a common datastore. Thus, distinct synchronization requests from the same Client can be routed to any physical Server in a load-balancing cluster if the Server can access the sync anchor corresponding to the Client's datastores. Different packages pertaining to the same synchronization session, however, are best routed to the physical Server to which the first package was routed.

Build a Secure Synchronization Platform

SyncML adopts a practical approach to security. It realizes the importance of security, but more importantly it realizes the tradeoffs between security, usability, and performance. It is clear that a single "one-size-fits-all" security solution is not likely to work, as SyncML application requirements are diverse. Day-to-day personal information management applications will use SyncML. For such applications, security requirements are often not very stringent. Enterprise applications will also use SyncML. For such applications, security requirements could be stringent.

To address these diverse requirements, SyncML allows the use of the Secure Sockets Layer (SSL) protocol for data security, but does not make the use of SSL mandatory. SyncML also allows multiple types of authentication. Clients can authenticate using a simple password mechanism. Clients can also authenticate using encrypted message digests, which are more secure.

SyncML also allows different granularities of authentication. Authentication could range from per Client, to per datastore, to per datastore object. Object-level authentication could be important in applications where different users can have different levels of rights to shared data. For example, a manager can share her calendar with her employees, so all can view and update public entries, but only the manager can view and update private entries.

SyncML does not require data encryption, as encryption can be prohibitive for certain mobile devices. Encryption is allowed, however, if Clients and Servers choose to encrypt the data they exchange across the network.

Build Upon Existing Web Technologies

Leveraging existing technologies and riding on established momentum is key to the success of any effort, especially standards efforts. SyncML chooses to use XML for its Representation Protocol not only for purely technical reasons but also to leverage the momentum around XML. By choosing XML, SyncML is readily able to use a wide array of tools including various parsers. SyncML leverages the MIME standard for data formats. The SyncML packages themselves conform to a registered MIME type.

The initial transport bindings chosen in SyncML are all established wide-area wireline, long-range wireless, or short-range wireless transport. HTTP is the underlying protocol for most of the World Wide Web. WSP is the emerging protocol for a wide class of WAP phones. OBEX is the data exchange protocol for emerging Bluetooth and Universal Serial Bus (USB) devices.

Build a *Working* Specification

One of SyncML's most important goals from the beginning was not only to design and draft a specification but also build a working reference implementation for part of the specification. The reference implementation was intended for concurrent release with the specification, as a testament to the technical soundness and completeness of the specification.

The SyncML Initiative put a tremendous amount of additional effort into the reference implementation. A C-language programming API was designed on which applications could be written. An actual framework was designed that could be implemented. Two transport bindings, HTTP and OBEX, were implemented. End-to-end demonstration scenarios were designed such that the implementation and the specification could be validated.

The reference implementation has catalyzed many development efforts around SyncML. It provides SyncML supporters a concrete starting point and enables quick application development. Although neither the API nor the reference implementation are actually part of the specification, they have played key roles in the growing acceptance of SyncML.

Promote Interoperability

The overriding goal of SyncML is interoperability. To that effect, SyncML has designed a detailed conformance test suite aimed at testing SyncML implementations for conformity to certain key aspects of SyncML. In addition to conformance tests, SyncML also defines a process by which interoperability is tested between conformant Clients and Servers. The SyncML Initiative regularly hosts "SyncFests," during which a product must show interoperability with two or more other products (from different companies) to be deemed an interoperable implementation. Chapter 13 covers SyncML interoperability testing in detail.

Architectural Insight into SyncML

SyncML adopts a layered view of the software architecture for mobile data synchronization, as depicted in Figure 4–2. This layered view is similar to a layered network protocol like TCP/IP but at a different semantic level. The overall synchronization system consists of the following layers:

- Application
- Data
- Synchronization
- Transport
- Physical

Architectural Insight into SyncML 73

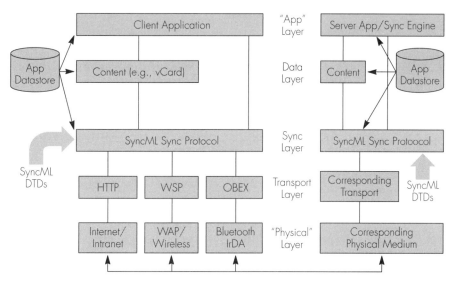

Figure 4–2
The layered architecture of the synchronization framework and the position of the SyncML Synchronization Protocol within the overall software stacks of mobile devices and network servers.

The application layer consists of the actual application, such as a calendar application, and associated synchronization logic, usually separated into a synchronization engine. There may be a single synchronization engine acting for many applications or one synchronization engine per application. At this layer, conceptually an application on a Client is communicating with an application on a Server. The communication and exchange of information actually occurs by crossing many vertical layers. At this layer, two synchronizing applications should focus on application semantics, such as what constitutes conflicting updates.

There is an underlying layer of data below most applications. At the data layer the peers are not two applications but two datastores. A memory-resident file on a mobile Client could correspond with entries or rows in a Server database. The data layer identifies the kind of data being synchronized. Calendar applications may use vCalendar data, contact applications may use vCard [VCARD21] data, and health-care applications may use patient records. Note that the format of the synchronized data may be different from the native format of the data stored in the two datastores in this layer. In most cases the native data is translated into the standard format during synchronization.

The synchronization layer constitutes the majority of SyncML specifications. This layer is concerned with the logical communication of changes to data items. The Synchronization Protocol determines the flow of the actual conversation between a Client and a Server during synchronization. The SyncML Representation Protocol DTD primarily determines the exact form of each communication. To effectively accomplish this, the synchronization layer also includes data-specific attributes, such as identification of datastores and identification of data items. Note that the synchronization layer is mostly concerned with the communication of changes and not their *interpretation*. The application layer determines what changes need to be communicated and the data layer determines the format of the data embedded in the communication.

The transport layer consists of the network transport protocols that are used. This layer is concerned with the actual messages that are exchanged, in contrast with the logical packages specified by the synchronization layer above. The actual form of one or more physical messages corresponding to a logical package could be different depending on the transport being used. The headers, encoding, and other aspects of messages exchanged over HTTP are different than those exchanged over WSP and OBEX. The physical layer consists of the actual communication medium being used, which can range from wide-area wired networks to local area networks, cellular networks, or even Bluetooth networks. One transport layer protocol can operate over multiple physical networks. For example, HTTP can operate over wired and wireless networks, while OBEX can operate over Bluetooth or IrDA links.

In summary, the various layers provide different levels of abstraction. Typically, an aspect of a higher layer can be realized by multiple entities in a lower layer. It is important to appreciate all the abstractions and software parts involved in end-to-end data synchronization to clearly understand the place and role of SyncML in data synchronization.

The Meaning of Synchronizing Two Datastores

What does it actually mean to synchronize two datastores? One possible answer is that after the process of synchronization, the two datastores are identical. This appears to be a limiting definition. For example, a mobile client device may store a limited subset of fields in a data item while the corresponding server data item stores all the fields. The mobile client device may also identify data items differently than a server. It may also be desirable for a mobile client only to store certain subsets of all data items in a server datastore. In addition to variations in the data items stored, synchronization has different meanings when different kinds of datastores are being synchronized. For example, adding a document or a data item to a datastore has different semantics than adding a transaction to a queue.

To be practical, SyncML must allow the above flexibilities. SyncML depends on a notion of "equivalence" between data items in two datastores that are being synchronized. Datastore A is synchronized with datastore B if and only if for every data item in store A there is a corresponding equivalent (but not necessarily identical) data item in datastore B. SyncML does not define the specific definition of equivalence that is appropriate for two particular datastores but requires that a *notion* of equivalence exist. For example, a calendar record in a mobile device can be deemed equivalent to a calendar record in the server, even though the one in the mobile device does not have information related to conferencing facilities in the meeting room pertinent to the calendar entry. Note that if datastore A is deemed synchronized with datastore B, this does not imply that datastore B is synchronized with datastore A. In other words, synchronization does not have to be reflexive. This allows datastores on Client devices to be subsets of datastores on Server devices and still be synchronized using SyncML.

Language of Synchronization: XML and MIME Usage

SyncML is primarily about expressing operations on data and communicating such operations in a structured manner. Since the operations are on certain data items, the data items themselves need to be expressed and communicated in a structured, consistent manner. SyncML defines various operations corresponding to adding, replacing, and deleting data items in a datastore. Such operations are performed on calendar data, contact data, relational data, XML documents, and various other kinds of data. SyncML uses the XML syntax for expressing operations. It uses MIME data types to identify data formats.

SyncML Messages are specified using well-formed XML. They do not need to be valid XML. This relaxation allows for terseness in SyncML Messages. SyncML makes heavy use of XML namespaces. Namespaces must be declared in the first element type that is used in a package from a particular namespace. SyncML also makes use of XML standard attributes such as "xml:lang". Any XML standard attribute can be used in a SyncML Message. The current SyncML DTD defines a namespace that is identified with a specific Uniform Resource Identifier (URI) and a specific Uniform Resource Name (URN).[9]

XML can be viewed as more verbose than alternate binary representations. This is often cited as a reason why XML (and SyncML) may not be suitable for low-bandwidth wireless networks. In most cases, SyncML uses shortened element type and attribute names, providing minor reductions in verbosity. Additionally, SyncML messages can be encoded in efficient tokenized binary formats such as WBXML. The use of the binary format is external to the specification and should be transparent to any SyncML application. The combination of shortened element type names and binary format makes SyncML competitive for usage over low-bandwidth networks. Clearly, the main advantage of XML is its international widespread acceptance for document markup. It allows for both human readability and machine processability. It allows one to capture document structure as well as content. This is extremely useful for data synchronization, where not just capturing content, but capturing structure semantics is essential. The overall benefits of using XML far outweigh the performance concerns.

SyncML leverages the MIME industry standard for specifying content types. Different data types in SyncML are expected to be registered MIME types such that a SyncML processor can determine how to process encapsulated data. SyncML messages themselves are registered MIME types. Clear text SyncML messages are registered as "application/vnd.syncml+xml", while binary encoded messages are registered as "application/vnd.syncml+wbxml". For transport-level protocols such as HTTP that support MIME content types, the MIME types for SyncML messages must be used.

---

9. The URI for SyncML DTD version 1.0 is http://www.syncml.org/docs/syncml_represent_v10_20001207.dtd. The URN is //SYNCML//DTD SyncML 1.0//EN.

5

Synchronization Protocol

To understand SyncML® technology and how different SyncML protocols and components interact with each other, it is best to understand the Synchronization Protocol [SSP02] first. The Synchronization Protocol uses the functionality from the other SyncML specifications, such as the Representation Protocol [SDS02], the Device Information DTD [SDI02], and the Meta Information DTD [SMI02]. The Synchronization Protocol defines how to use those protocols and DTDs in a consistent and interoperable manner. In general, the purpose of the Synchronization Protocol is to define the interaction between the Client and the Server, or rather, the phases for accomplishing a complete SyncML synchronization session.

By mastering the Synchronization Protocol, it is easy to go forward and learn more about other SyncML components. Thus, it is useful and important to read this chapter carefully even though your technical interests may be other than the Synchronization Protocol.

Overview

The SyncML Initiative defined the Synchronization Protocol because simply providing DTDs (i.e., the syntax of messages) did not do enough to enable synchronization between devices. First, full end-to-end interoperability cannot be achieved only by defining the DTDs, as they give too much flexibility. Second, several functions, such as the authentication procedure, need to be defined by the Synchronization Protocol so consistent synchronization can really take place.

By following the Synchronization Protocol, implementations know which DTD element types they can use and when they can use them. In addition, the protocol adds extra functionality to the SyncML technology through combinations of element types defined by the DTDs.

The design of the Synchronization Protocol was based on common synchronization scenarios, or use cases from the end-user point of view. Such common synchronization scenarios include:

- Two-way synchronization between a device and a server
- Back-up and restore operations between a device and a server

Relation to the Representation Protocol and Other DTDs

The Representation Protocol defines the format of SyncML Messages, is independent of the other SyncML specifications, and can be used for any synchronization scenario. The Synchronization Protocol is completely different. The Synchronization Protocol is dependent on the format and DTD element types of the Representation Protocol. Also, the Synchronization Protocol is targeted at specific synchronization scenarios.

The two other DTDs, the Meta Information DTD (the MetInf DTD) and the Device Information DTD (the DevInf DTD) are tools from the Synchronization Protocol point of view. The MetInf DTD is used for complementing the Representation Protocol when creating completely meaningful SyncML Messages. On the other hand, the Synchronization Protocol utilizes the DevInf DTD to provide basic service discovery functionality within the SyncML technology. Although the MetInf DTD and DevInf DTDs can be utilized outside the SyncML technology, the Synchronization Protocol links them closely to it and uses their element types inside the SyncML Message structure.

Figure 5–1 depicts the relation of the Synchronization Protocol to the other SyncML components when creating a logical output from a logical input. The input can be either an internal one (e.g., the user interaction) or an external one (e.g. a SyncML Message). Likewise, the logical output can be internal or external.

Entities Using the Synchronization Protocol

The Synchronization Protocol is not symmetrical, although it is close. Because of this, it is useful to carefully watch the roles defined by the Synchronization Protocol. The roles have a strong impact on implementations and on the SyncML features that need to be supported. Figure 5–2

Overview

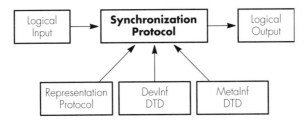

Figure 5–1
Relation of Synchronization Protocol to other components

Figure 5–2
Entities using Synchronization Protocol

shows that the Synchronization Protocol is used between two entities, a SyncML Client (hereafter, Client) and a SyncML Server (hereafter, Server). The Synchronization Protocol specification calls them Device Roles.

When two entities are using the Synchronization Protocol, one of them must take the Client role and another the Server role. However, this does not mean that two servers or two devices could not communicate with each other using this protocol. To do so, one has to temporarily take the Client role or the Server role for enabling the communication. The Synchronization Protocol may not be optimized for this kind of synchronization but there are no restrictions that prevent this.

The Client role is usually selected for mobile devices such as mobile phones or PDAs. In some situations, the Client role can also be taken by a PC if the PC synchronizes data with a Server. The Client's functionality is much simpler than the Server's because there are fewer features to be supported at the protocol level, as well on the application level. Basically, the Client is responsible for collecting data to be sent to the Server, sending it, receiving data from the Server, and storing it. However, there are no definite requirements for the Client to analyze payload data itself. Payload is the actual data being synchronized, e.g. contact or calendar data.

As a result of the simpler mandatory Client functionality, the memory and footprint requirements are kept to a feasible level. This is very important when considering that SyncML is seen in mass volume products, which, due to cost optimization, cannot provide much extra memory or processing power. SyncML technology does allow for extra functionality in more sophisticated Clients, such as PCs or advanced PDAs.

The Server role is definitely more complex than the Client one. Naturally, the desired functionality and quality level affects how complicated the Server implementation needs to be. In addition, the desired Server environment has an impact on the implementation complexity. For instance, an Internet Server handling a large number of Clients has a quite different architecture than a PC implementation synchronizing with only one device at a time.

Supported Synchronization Scenarios

The Synchronization Protocol defines a set of synchronization scenarios to be implemented. They are called Sync Types. The Sync Types are specified to provide different types of synchronization. These Sync Types are described in Table 5–1.

Table 5–1
Sync Types

Sync Type	Description
Two-way sync	A common Sync Type in which the Client and the Server exchange information about modified data in these devices.
Slow sync	A form of two-way sync in which all items are compared with each other. The Client sends all its data from a datastore to the Server and the Server does the sync analysis for this data and the data in the Server to determine the delta between the Client and the Server, which then needs to be communicated to the Client.
One-way sync from Client only	A Sync Type in which only the Client sends its modifications to the Server.
Refresh sync from Client only	A Sync Type in which only the Client sends all its data from a datastore to the Server. From the Client perspective, this is commonly called the export.

Table 5–1
Sync Types (Continued)

Sync Type	Description
One-way sync from Server only	A Sync Type in which only the Client gets all modifications from the Server.
Refresh sync from Server only	A Sync Type in which only the Server sends all its data from a datastore to the Client. This is commonly called the import operation from the Client perspective.
Server alerted sync	A Sync Type in which the Server alerts the Client to perform sync.

The Sync Types above can be categorized according to the direction in which payload data is sent. There are Sync Types in which payload data is sent in both directions. There are also Sync Types in which payload data transferred only in one direction, either to the Server or to the Client. One of the Sync Types, the Server alerted sync, does not really belong to any of these groups, as it is only a trigger for synchronization. The four separate Sync Type categories are described in Table 5–2.

Table 5–2
Sync Type Categories

Category	Description	Included Sync Types
Bi-directional data transfer	Data transfer from the Client to the Server and vice versa is provided.	Two-way sync, Slow sync
Unidirectional data transfer to Server	The data can be transferred from the Client to the Server only.	One-way sync from Client only, Refresh sync from Client only
Unidirectional data transfer to Client	The data can be transferred from the Server to the Client only.	One-way sync from Server only, Refresh sync from Server only
Trigger, no payload data transfer	The Server can initiate synchronization by triggering the Client.	Server alerted sync

It is worth noting that the last category (the Trigger) can be used to initiate any of the Sync Types belonging to the other categories. For example, by using the Server alerted sync the Server could send a trigger for the One-way sync from Server only.

Phases of Synchronization Protocol

The Synchronization Protocol consists of distinct logical phases or handshakes. The phases together comprise a complete synchronization session. If the Client initiates a synchronization session, the first phase is the Initialization phase. This phase is also introduced as a common beginning stage for all Sync Types in the Synchronization Protocol. The second phase, the Data Exchange phase, is dedicated to transferring modifications between a Client and a Server. The direction of the transfer is dependent on the Sync Type used. The last phase during a SyncML synchronization session, called the Completion phase, is needed to finalize the synchronization session.

The Server alerted Sync Type can occur prior to the Initialization phase and it can also be categorized as a fourth phase. As explained earlier, it is a trigger for other Sync Types from the Server side. This phase is optional and it is only run when the Server initiates a synchronization session.

The Synchronization Protocol specifies numbered Packages that are associated with different phases. Figure 5–3 depicts a SyncML synchronization session between the Client and the Server in the form of a Package sequence chart.

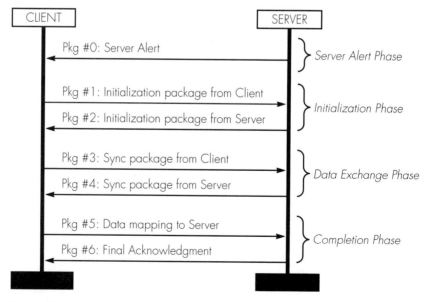

Figure 5–3
Numbered Packages in a SyncML Session

One SyncML Package can include multiple SyncML Messages. Due to this, a Package with a lower number may not be completely sent before a Package with a higher number is received. However, the Synchronization Protocol defines which commands can be sent in a later Package before the previous Package has been completely received. Thus, it is not a problem that a Package can include multiple Messages.

Due to the existence of the phases and the requirement for sequencing them, the Synchronization Protocol can be modeled as a state machine. Naturally, this also means that the implementations utilizing this Protocol cannot really be stateless either in the Client or the Server end.

Initialization

The Initialization phase is mandatory for any synchronization. Vital functions are performed during Initialization. These functions include:

- Authentication between the Client and the Server
- Exchange of the device and service capabilities
- Indication of synchronized content and the Sync Type
- Conformity check of a previous synchronization session

The Initialization phase is analogous to the logon procedure to a server or a service. It may simultaneously occur with the Data Exchange phase if desired by the SyncML Client. In practice, most implementations do the Initialization phase separately from the Data Exchange phase.

Authentication

Authentication between the Client and the Server can be bidirectional. The Client can authenticate itself to the Server and vice versa. SyncML does not mandate bidirectional authentication and allows one party to waive authentication of the other party. Typically, only the Client authenticates itself to the Server, unless it is using the Server alerted Sync Type. If the Client does not automatically authenticate itself, the Server usually challenges the Client to do so. If the Client does not authenticate, and the Server requires authentication, then synchronization is not continued.

The SyncML authentication procedure is well defined in the Synchronization Protocol specification, but there are a few critical issues which developers should be aware of when implementing SyncML:

- Session- or message-based authentication
- MD5 digest authentication and handling nonce
- First synchronization between the Client and the Server

It is worth noting that authentication can be accepted for an entire SyncML session or only to authorize a given SyncML Message. In the first case, the authentication takes place once in the Initialization phase. In the latter case, it is done for each Message. In general, since the Server requires authentication of the Client, the Server has the power to decide which scheme is selected. Both schemes have different characteristics, as well as benefits and drawbacks. Every implementation needs to make a decision as to which one is utilized.

A positive characteristic associated with the message-based authentication scheme is a higher security level in general. In the case of Basic authentication, the message-based authentication scheme can reveal the password if SyncML is used over a nonencrypted transport. Also, the message-based authentication scheme can offer benefits if the communication is routed to a different Server in the middle of a synchronization session. A benefit of using the session-based authentication scheme is more optimized bandwidth consumption, as less data needs to be transferred. In addition, session-based authentication decreases processing requirements, as the authentication is only done once.

When utilizing the MD5 digest authentication, a nonce[1] is transferred between a Client and a Server, which is not the case in Basic authentication. The nonce is typically saved for the next session, but this does not have to be the case, and implementations should be aware of that. If it is not stored for the next SyncML session, the implementations challenging for the MD5 digest authentication need to be prepared to send the nonce again at the beginning of the following SyncML session. This applies whether the authentication scheme is message-based or session-based.

When doing the first synchronization between a Client and a Server, the credentials based on the Basic authentication should not be sent in the first Message. The reason is that neither the Client nor the Server can know which authentication procedure, Basic or MD5 digest authentication, is desired by the Server or the Client, respectively. Sending the credentials based on the Basic authentication in the first Message causes a risk that the security benefit gained by later using the MD5 digest authentication will be lost, as the credentials have already

---

1. A nonce is a uniquely generated data string that creates a digest value when responding to the challenge of an MD5 digest authentication.

been revealed in the first Message. Actually, when doing the first synchronization between a Client and a Server, the Client should not send any credentials to the Server at all in the first Message. The Client should await the authentication challenge in the Server's response to the first Message. After that, the Client knows which authentication type it needs to use with the Server. The same behavior applies to the Server.

Exchange of Device Capabilities

The purpose of exchanging device capabilities is to inform synchronizing entities about characteristics related to the software and hardware of the entities, as well as services offered by these entities. In principle, this information helps the entities ascertain what they can synchronize with each other and how they can synchronize. In other words, the exchange of device capabilities is similar to a service discovery mechanism. For example, the entities can find out the following characteristics related to a device or a service:

- The manufacturer, model, and version
- The supported applications (e.g. a calendar application)
- The datastore addresses and data formats (e.g. vCalendar) associated with the supported application
- The supported Sync Types (e.g. Two-way sync)

The exchange of device and service capabilities is especially crucial to the Server. That is, the Server needs the device capabilities of a Client in order to offer a high-quality synchronization service to the Client. This way, the service can be operational and customized according to the Client capabilities. For the Client, the device capabilities of the Server may not be so vital, because many embedded devices are not able to process that information. However, there are also Client products that do require the device capabilities from the Server. The details of the device and service capabilities are considered in Chapter 7.

At the Protocol level, the exchange of device capabilities is accomplished by using the Get and Put operations as defined by the Representation Protocol. This means that a Client or a Server can send its own device capabilities and request the capabilities of the other entity. The device capabilities are encoded as defined in the Device Information DTD specification.

Indication of Synchronized Content and Sync Type

When synchronizing data between a Client and a Server, it is also important to know which datastores are to be synchronized. This information is exchanged during the Initialization phase. The datastores are

```
                    <Alert>
                    <CmdID>1</CmdID>
   Sync Type    {   <Data>200</Data>
   Definition

     DB URI     {   <Item>
   Definitions      <Target><LocURI>./server_db</LocURI></Target>
                    <Source><LocURI>./client_db</LocURI></Source>

                    <Meta>
                    <Anchor xmlns='syncml:metinf'>
  Sync Anchor   {   <Last>234</Last>
   Definitions      <Next>276</Next>
                    </Anchor>
                    </Meta>

                    </Item>
                    </Alert>
```

Figure 5–4
Example of Alert command in Initialization[1]

1. *The XML encoding is used in the example to make it more readable. Different encodings are discussed in the following chapters.*

indicated by specifying Uniform Resource Identifiers (URIs) for datastores. Similarly, there is also a need to exchange information on how data is synchronized. This is done by indicating the Sync Type.

To specify all this information, the Alert command defined by the Representation Protocol is exchanged between a Client and a Server. Within an Alert command, the datastore URI and Sync Type are indicated. Figure 5–4 shows an example of an Alert command used in the Initialization Package from a Client to a Server. The figure shows how a Sync Type and the URIs are specified within an Alert command.

Conformity check of previous synchronization

In the Initialization phase, the Client and the Server are also able to check that they agree on a marker that indicates when they last synchronized. In other words, the Client and the Server can ensure that they have a common understanding of when the previous synchronization took place. The marker can be a timestamp or a sequence number. Comparing the markers enables an entity to detect if something has failed after the previous synchronization time. The reasons for failure may be various. For example, the devices may never have been synchronized with each other, the device may have completely been reset after a previous synchronization time, or a previous synchronization may have failed in the middle of the process. The last case should not usually cause a failure in the conformity check but might do so if an implementation is not well designed.

The Synchronization Protocol uses the sync anchors of datastores to enable a conformity check. In other words, the sync anchors are the markers. They are specified within an Alert command as depicted in Figure 5-4. A receiving entity needs to be able to store the anchor called Next such that the conformity check can be processed during the following synchronization session. The Server implementations are required to do this, but it is optional for the Client implementations, because it is assumed that the Server implementations are more robust.

Naturally, there has to be a way to recover if the conformity check fails. A Server usually initiates a Slow sync with a Client to recover from this situation. This can cause a lot of data to be exchanged, and synchronization can take a long time. This kind of situation should be avoided if possible. Wireless connections can often be interrupted unpredictably. Consequently, synchronization sessions will also often fail, as the normal end of the session will not be accomplished. Due to this, the timing of updating the sync anchors is essential. Having the update at the very end of a synchronization session can dramatically help to avoid unnecessary slow syncs.

Example of Sync Anchor Usage. Figure 5-5 depicts an example of Sync Anchor usage on the Server side. The Client sends the anchors to the Server and the Server stores and analyzes them. In this example, a Client and a Server synchronize twice with each other (Session #1 and Session #2). After Session #1 the persistent memory of the Client is reset. As a result, at the beginning of Session #2, the last sync anchor of the Client does not match the stored sync anchor of the Server. The Server detects this, and sends a slow synchronization alert to the Client. Typically, a Client will begin a slow synchronization with the Server upon receiving this alert.

In this example, Session #1 was started on 07/06/2002 at 08:09 AM. The previous successful session with the Server was started on 07/06/2002 at 03:04 AM. At this synchronization, slow synchronization is not initiated because the last sync anchor sent by the Client matches the last sync anchor that the Server had stored for this Client. Session #2 was started on 07/07/2002 at 01:02 AM. Due to a memory reset between Session #1 and Session #2, the last Sync Anchor is empty. The

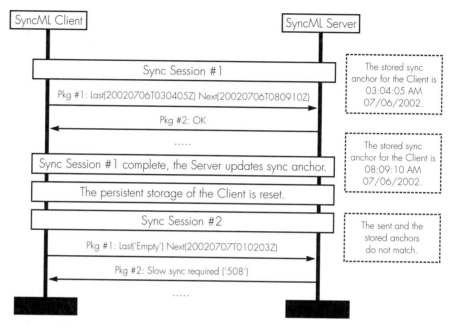

Figure 5–5
Example of Sync Anchor usage

last sync anchor sent by the Client does not match with the last sync anchor stored by the Server (07/06/2002 at 08:09 AM). The Server therefore alerts the Client to initiate a slow synchronization.

Data Exchange

The primary goal of the Data Exchange phase is to provide an interaction for transferring all modifications that have happened since a previous synchronization. The modifications can include payload data—i.e., data to be synchronized. An Add operation is one type of modification for which payload data needs to be included. It is not always necessary to transfer any payload data in order to exchange a modification between synchronizing entities. A Delete command represents this type of modification.

The direction in which modifications are sent is dependent on which Sync Type is being used. In addition to that, the direction is also dependent on whether any modifications have been done to the entities involved since the last synchronization. This means that although a Two-way Sync Type is utilized, modifications will not be sent from an entity to another if there were no modifications since the previous synchronization time.

Sometimes, when transferring modifications from a Client to a Server, not all modifications including payload data can be sent in a single SyncML Message. Instead, a SyncML Package includes multiple SyncML Messages. In the case of sending Client modifications, the Server waits until the Client has sent all modifications before the Server starts to send any of its modifications to the Client. This way, the Server has the potential to detect all possible conflicts. A conflict means here that the same data item has been modified in both the Server and the Client since the previous synchronization time.

Completion

The Completion phase is designed to end a session properly and ensure that the synchronizing entities have received all the information they will need to synchronize with each other again later on. The main functions of the Completion phase are the following:

- The Client reports how it has succeeded in processing the modifications from the Server.
- The Client sends the ID mapping information for all added items back to the Server.
- The Server informs the Client that it has completed everything.

It is mandatory to support the Completion phase, although this phase might not always take place. A synchronization session can be either completed after Package #4 or Package #6. If the Server does not sent any modifications to the Client and it indicates that it does not require a status report as a response to Package #4, then the session is ended after Package #4.

When synchronization includes the transfer of modifications from the Server to the Client, the Client needs to respond to the modifications by sending status information back to the Server. The Server can explicitly waive this by informing the Client that it does not want to get the responses to these modifications. This status information is included in Package #5.

In Package #5, the possible identifier mapping information is also included. The mapping information is needed if the Server has used one or more Add operations. The need for the mapping operation is based on the fact that Clients and Servers commonly use different identifiers for the data items. By sending the mapping information to the

Server, the Client ensures that the Server knows which identifier(s) the Client uses for the data item(s) added by the Server. By knowing this, the Server can use the Client IDs when addressing items on the Client.

After sending all the mapping information to the Server, the Client waits to receive the final acknowledgment from the Server. The acknowledgement is delivered by sending the status information for the mapping operations. This status information is included in Package #6. After receiving Package #6, the Client knows that the synchronization session is over.

For the Server, the definition of the very end of a session is more difficult because the Server sends the last Package. Thus, the Server cannot be sure on the SyncML level whether the Client has received the last Package or not. However, some lower-layer transport protocols provide capabilities that can offer the ability to check whether the data transfer has been completed successfully. If this kind of functionality is available, the Server can use this information and update the sync anchors when the data transfer has completed successfully. If this kind of functionality is not available, the Server needs to update the sync anchors after sending the last Package to the Client.

Server Alert

The Server Alert phase cannot really be called a handshake, because only one Package is sent from a Server to a Client. However, the status information related to this Package and the commands inside is sent in Package #1. Thus, this phase can also be thought of as a preinitialization.

The main purpose of the Server Alert phase is to indicate to the Client that it should start to synchronize with the Server. The details of the Server Alert phase are very similar to the Initialization phase. The Server Alert serves the following purposes:

- Authentication between the Client and the Server
- Indication of synchronized content and a Sync Type

These functions are also present in the Initialization phase. However, the exchange of device capabilities and the conformity check of a previous synchronization session are not done as part of the Server Alert phase.

Authentication is relevant in the Server Alert phase because the general statement that only the Client usually authenticates itself to the

Server is not necessarily true here. In this case, the server authentication should prevent a solicitation to synchronize with a hostile Server.

An example of the hostile Server could be a Spying Server, which attempts to get all confidential calendar information from a confidential Client. The Spying Server could initiate a synchronization session for calendar data. If the Client did not require any authentication from the Spying Server, the Server could easily download all the information from the Client. As a result, the Client would not be so secure or confidential.

To summarize, the Client implementations supporting the Server alerted sync should carefully consider whether to accept the alert from a Server without credentials. At the very least, some sort of user acceptance should be involved if no server authentication is implemented or desired at the Client end.

The second main function of the Server Alert phase is to indicate which kind of data (i.e. content) is synchronized and which Sync Type is to be used for that data. This function is similar to the procedure done in the Initialization phase when the Client indicates this type of information to the Server.

Transferring Large Amounts of Data

The Synchronization Protocol allows the transfer of large numbers of modifications or large payload objects when there are device limitations relating to the size of SyncML Messages. To achieve this, each SyncML Package can be divided into multiple SyncML Messages. This allows the synchronization of a large number of modifications or large payload data without exceeding the maximum size of a SyncML Message.

Large Object Delivery

The SyncML 1.0.1 specification does not include the functionality to divide a large payload object, such as an image file, into multiple Messages. That is, if a large data object within a SyncML command is transferred by utilizing the SyncML 1.0.1 specification, the command, including the data object, needs to fit into one SyncML Message. When considering the fact that the sizes of SyncML Messages range between a few kilobytes and tens of kilobytes, the overall size of the large object cannot be very large. The version 1.1 (and later) SyncML specification takes the delivery of large objects into account and provides the functionality for transferring data objects with sizes larger than the maximum size of a SyncML Message.

From the Synchronization Protocol point of view, the functionality for transferring large objects over multiple Messages means that a segmentation and reassembly feature is provided inside SyncML commands such as Add and Replace. This is like any segmentation and reassembly feature provided by other transport protocols, but now it can be used inside individual SyncML commands. For example, if a Server wants to add a large picture to the image database of a Client, but the size of the image is larger than the Message size that can be received by the Client, the Server will distribute the image over multiple SyncML Messages.

Maximum Size of SyncML Messages

Before analyzing the features for transferring large numbers of modifications or large payload objects, it is worth considering the factors that affect the maximum size of a SyncML Message. The Client and the Server can define how large a SyncML Message they can receive and send. Any or all of the following reasons may ignore the size of a SyncML Message:

- Internal characteristics of a device
- Transport protocol for encapsulating SyncML Messages
- Properties of a gateway to access a network
- Underlying physical networks

The first reason in the list is quite obvious when considering that SyncML devices can vary from small wireless embedded devices to large enterprise servers. Commonly, an internal characteristic affecting the maximum size is the amount of RAM (Random Access Memory) available for buffering a SyncML Message. In a small wireless device, the RAM available for buffering is usually on the order of a few tens of kilobytes, and rarely more. There are other characteristics as well, but RAM limits may be the most compelling one in wireless devices.

The transport protocol carrying SyncML Messages can also have an impact on the maximum size of a Message. This is the case if the transport protocol does not offer the functionality for chunking data over a network. This means that a SyncML Message can be transferred in small pieces and reassembled at the destination. Not all protocols offer this kind of functionality; some are only able to transfer complete

SyncML Messages. One example of this kind of protocol is the Wireless Session Protocol (WSP) [WSP01] defined by the WAP Forum®.

The third factor affecting the maximum size can be the gateway used for accessing a network. A WAP gateway (interconnecting a wireless network and a fixed network) is a good example of a gateway that limits the maximum size. When a wireless device connects over WAP to the Internet, the limitations of the WAP gateway can be more restrictive than the internal characteristics of the device. The maximum size in such a situation is determined by the characteristics of the WAP gateway, not the internal limitations of the wireless device.

The underlying physical network can have design-oriented issues impacting the maximum size of a Message. In other words, larger-size Messages can affect the robustness and performance of the system. When implementing SyncML on devices enabling connections over fixed networks or wireless networks, the fundamental characteristics of these networks, such as data rate and latency, must be taken into account.

Multiple Messages per Package

A Client or a Server will need to divide a large amount of modifications into multiple Messages if they do not fit into one Message. For example, a Server has fifty modifications to be sent to a Client. However, the maximum size of a SyncML Message that can be received by the Client would be exceeded if the Server embedded all the modifications into one SyncML Message. Thus, the Server must distribute the modifications over multiple Messages.

As defined in the Representation Protocol, specifying a Final tag inside a Message indicates the end of a SyncML Package. If the tag is not specified, a Package is not completed yet and more Messages for the same Package need to be received.

The Synchronization Protocol is designed so that if a device receives a Message without a Final tag, it needs to ask for a next Message, explicitly or implicitly. An explicit request means that the device that received the incomplete Package sends a special alert. An implicit request is done by sending the status information related to the Message or to commands within the Message. It is possible to use both ways simultaneously. The implicit way alone may be more common and recommended, as devices in general require the status information to be sent. Thus, the implicit request often comes automatically, in which case the explicit request would only be redundant. Naturally, if a device

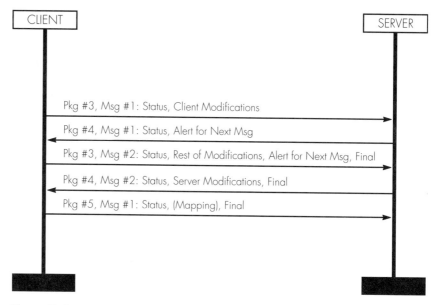

Figure 5–6
Example of using multiple Messages per Package

does not request any response to a Message without a `Final` tag, the explicit request mechanism needs to be used.

Figure 5–6 shows an example of using multiple Messages. In this example, Package #3 is distributed over multiple Messages. In the figure, it is shown that Package #4 is divided into two Messages, as all the modifications from the Server to the Client cannot be included in the first Message belonging to Package #4. The example uses both the implicit and explicit ways for requesting more Messages.

Mapping of Identifiers and Slow Synchronization

In this chapter, handshakes and other features of the Synchronization Protocol have been introduced, and impacts have been considered. All of them together create the overall functionality provided by the SyncML Synchronization Protocol. Now is a good time to closely look at two important features of the Synchronization Protocol: identifier mapping and slow synchronization. Their impact on implementations and the overall infrastructure is explored below.

Nature of Identifier Mapping

As explained earlier, the whole idea of identifier mapping is to enable the Client and the Server to address data items using distinct identifiers. When synchronizing data and exchanging modifications, the identifiers allocated by a Client are used. The Server identifiers are only used when the Server adds a data item to a Client. Because of this, the Server needs to maintain identifier mapping with every Client that it synchronizes with. From the Client standpoint, this is a huge benefit because the Client does not have to store the long globally unique identifiers (GUIDs) used by the Servers.[2] For wireless devices implementing the Client functionality, using the same data identifiers as the Server would, in general, be an impossible requirement.

Figure 5-7 depicts a simple situation in which one Server (Server A) only synchronizes with two Clients (Client A and Client B). Now, Server A has to keep two mapping tables, one for Client A and one for Client B. In other words, the Server knows which Client ID needs to be used for each data object having the mapping information. Also, if the Clients refer to a data item with a LUID (local unique identifier), the Server knows at which data item a modification is targeted.

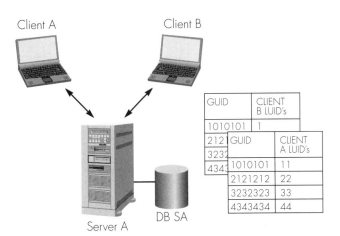

Figure 5-7
One Server and two Clients in synchronization environment

---

2. The Servers commonly use globally unique identifiers (GUIDs) for data objects. The lengths of those GUIDs are typically in the range of 64–128 bytes. In practice, Clients use locally unique identifiers (LUIDs), whose lengths are equal to or less than 16 bytes.

Figure 5–8 depicts a more complicated situation, in which two Servers (Server A and Server B) synchronize with two Clients (Client A and Client B). Both Server A and Server B have the ID mapping of Clients A and B. It is possible that duplicate items may occur. For example, if Client A and Client B (both containing the same item) synchronize with Server A, duplication may occur due to the fact that the Clients have different LUIDs. Since the Client sends the modifications first, the Server has the role of detecting if a data item already exists with a different identifier. In other words, the Server must have the ability to prevent the creation of duplicates

To understand this issue more, the example in Figure 5–9 reveals some points. In that example, a user adds a data item into Client A, which is then synchronized to Server A. After that, Server A and Client B synchronize with each other. As a consequence, the data item is now also found on Client B. It is then the turn of Client B and Server B to synchronize their data. Now, Server B has the data item. Finally, Client A and Server B synchronize. Client A obviously sends the data item to Server B because it does not know that the item already exists there. At this moment, Server B cannot count on ID mapping because it has no mapping for this item with Client A. Thus, it needs to use another mechanism to find out that it already has the data item. If this mechanism succeeds, no duplicate of this data item is created in Client A and Server B.

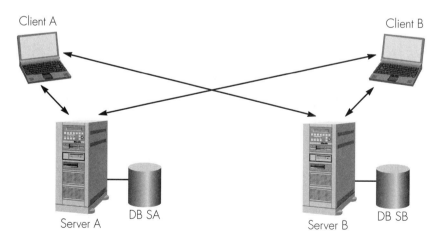

Figure 5–8
Two Servers and two Clients in synchronization environment

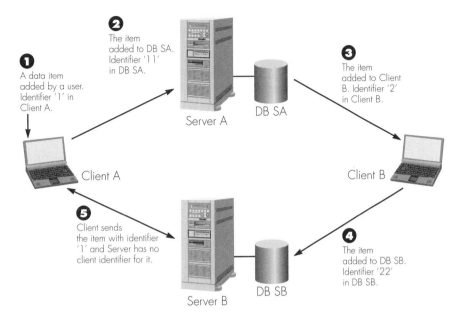

Figure 5–9
Example of potential duplicates in identifier mapping

Mechanisms for preventing duplicates are very much implementation-specific and, as such, are outside the scope of the Synchronization Protocol. Overall, those mechanisms are based on the analysis of the content of a data item. Basically, this means that the content of an incoming data item is compared to existing content before deciding whether a data item needs to be added.

Slow Synchronization

The way to recover from failures that may happen during synchronization is to use slow synchronization. It is also clear that this should be avoided if possible, since a slow sync requires the transmission of a lot of data. Nevertheless, slow synchronization is sometimes needed.

The Synchronization Protocol allows a great deal of flexibility regarding the functionality of slow synchronization. The Synchronization Protocol specification defines it in the following way: *"The slow sync is a form of the two-way synchronization in which all items in one or more databases are compared with each other on a field-by-field basis. In practice, the slow sync means that the client sends all its data in a database to the server and the server does the sync analysis (field-by-field) for this data and the data in the server. After the sync analysis, the server returns all needed modifications back to the client."*

Server implementations can quite freely decide how they process slow synchronization, as the specification does not really define it strictly. In practice, there are major differences in how implementations behave when a slow sync is initiated. The drawback related to this is inconsistency—i.e., the end-user experience offered by different servers can be very different. As a consequence, an optimized implementation of the slow sync operation is a good opportunity for a Server vendor to really differentiate itself and really show how good its implementation is. For instance, the differentiation can be related to the performance and the number of created duplicates.

Identifier mapping and slow synchronization have common elements because slow synchronization is always used when synchronization is done between a Client and a Server for the first time. It is possible that many new data items may enter the Server from the Client. Figure 5–10 gives an illustrative example to see how the slow synchronization operation links to the identifier mapping. In this example, Client A first synchronizes with Server A. After that, the Servers are synchronized with each other. Thus, the content, which was synchronized from Client A to Server A, is now in Server B, too. If it is assumed that the Client A and Server B were not earlier synchronized with each

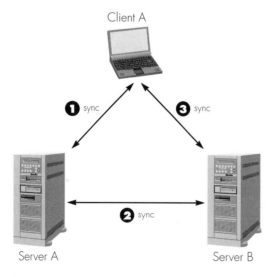

Figure 5–10
Slow synchronization in an environment of multiple devices

other, then slow synchronization is initiated when the two entities are synchronized with each other. When slow synchronization is started between Client A and Server B, all data items from Client A are sent. Server B should be able to detect that they exist already; if so, it will only need to update the mapping for these data items.

To summarize this subsection, the main messages to the developers are:

- The slow synchronization operation is a very powerful tool to recover from the failures, but it should only be used when really necessary.
- The identifier mapping operation is very useful when dealing with data items that have been synchronized earlier.
- When using the slow synchronization operation or adding new items, checking the existence of the items needs to be based on the content of the items.

By following these rules, Client and Server implementation can skip many quality- and performance-related problems, which will put them well on the way to providing a great user experience.

An Example Synchronization Dataflow

This section illustrates a hypothetical synchronization scenario between a Client and a Server using the SyncML Representation and Synchronization Protocols. The example shows a Two-way synchronization, in which both the Client and the Server exchange respective updates. For simplicity, possible failure scenarios such as unmatched synchronization anchors are ignored. For brevity, exchanged information is logically expressed, instead of illustrating exact syntax. The first scenario shows the changes made to data items on the Client and the Server. Then, each subsection of this section outlines how information is exchanged during synchronization and what actions the Server and the Client take. This is analogous to the Message sequence chart in Figure 5-3. This is a normal Two-way synchronization and the Client initiates the synchronization session. Therefore Package #0 is not needed in this example.

In this example, the Client has made the following changes to its datastore since the last synchronization time with this Server:

- Updating entry A
- Deleting entry B
- Inserting new entry C

During the same time the server datastore makes the following changes concurrently:
- Updating entry A
- Updating entry B
- Inserting new entry D
- Updating entry E

This will result in the following conflicts and the assumed corresponding resolutions:
- Update/Update conflict on entry A. Resolved in favor of the Server by the Server.
- Delete/Update conflict on entry B. Resolved in favor of the Client by the user. In other words, the Client deletes the update entry B coming from the Server.

Package #1: Client Initialization

This is the first Package (Package #1) in a Two-way synchronization used by the Client to initiate the session with the Server. The Package consists of the following:
- Header (with `serverID`, `clientID`, and authentication data)
- `Alert` command (with `type=Two-way-sync`, target datastore, source datastore, `Last` and `Next` synchronization anchors)

The header contains essential identifying information and authentication information. The `Alert` command indicates the requested Sync Type, and identifies the target and source datastores and the synchronization anchors.

Upon reception of this Message, the Server processes the header information, authenticates the Client, and authorizes it for synchronization with the target datastore. It compares the `Last` synchronization anchor with the stored `Next` synchronization anchor for this Client, identified by its `clientID`. The Server may inspect the Client's capabilities to make a decision if multiple Messages are needed for certain synchronization Packages in order to limit the buffering space needed for the Client to process the Packages.

Package #2: Server Initialization

This Package (Package #2) consists of the following:
- Header (with `serverID`, `clientID`, and authentication data)
- `Status` command for previous Client commands
- `Alert` command for Server synchronization anchors

The Server sends its authentication data to enable a Client to authenticate the Server, if necessary. This type of Server authentication is essential for applications where data on Client devices is sensitive or proprietary data that must be secured from theft. The Server may send its synchronization anchors which the Client may match with its synchronization anchors to determine if it should proceed with normal synchronization.

Package #3: Client Modifications

This Package consists of the following:
- Header
- Several Status commands to reply to the commands from Server
- Sync command with the following operations: Replace A, Delete B, and Add C

The Server performs the necessary authentication and begins to process the Sync command. It finds that the Client's update of entry A conflicts with its own update of the same item. The Server determines that its update takes precedence over the Client's update. It prepares a Status command (Status 1), indicating the conflict and its resolution, which is included in the next Package from the Server to the Client. It also finds that the Client's delete operation on item B conflicts with its update operation on the entry B. In this case, however, it determines that the Client should resolve this conflict. The Server prepares a Status command (Status 2), which indicates the identification of this conflict but no resolution. The Server proceeds to add the entry C to its datastore and prepares a Status command (Status 3), which indicates the addition of this entry. Since the entry C is a new item created by the Client, it only has a LUID associated with it. The Server determines the appropriate GUID for the entry C and updates its mapping table for this particular Client, storing the LUID and GUID for the entry C.[3]

Package #4: Server Modifications

This Package (Package #4) consists of the following:
- Header
- Status commands (Status 1, 2, and 3)
- Sync command with the following operations: Replace A, Replace B, Add D, and Replace E

---

3. See data identifier mapping details, discussed earlier in this chapter.

The `Status` commands include the status information described in the previous section. As discussed, `Status` 1 and `Status` 2 indicate the conflicts in entries A and B. The `Sync` command contains the Server's updates to entries A and B. For entry B, the Client ignores the command and keeps entry B as deleted. The Client also proceeds to add the new item D. The Client assigns the LUID for item D. The Client then creates a `Map` command, which contains the assigned LUID. Also, it prepares a `Status` command, which indicates the addition of this entry. The `Status` and `Map` commands are included in the next Package to the Server. The Client processes the `Replace` command for E, and makes the necessary update on its local datastore and also prepares a `Status` command for this update.

Package #5: Status and Map

This Package consists of the following:
- Header
- `Status` commands for the addition and update
- `Map` command for ID mapping

The `Status` commands include the status information described in the previous section. The `Map` command contains the Client's LUID and the Server's GUID for the new entry D. The Server extracts the LUID from the `Map` command and updates its data-identifier mapping table for the entry D. At this point (or just after sending Package #6), the Server also updates the stored sync anchor for the Client. This stored sync anchor must match the Client's `Last` sync anchor in the next session.

Package #6: Map Acknowledge

This Package consists of the following:
- Header
- `Status` command for the Client `Map` command

This Package primarily acts as a final acknowledgement from the Server that the synchronization is complete, including data-identifier mapping. At this point the Client may update any sync anchors that it may store for the Server.

6
Representation Protocol

The SyncML® Representation Protocol [SRP02] defines the syntax of information exchange between two synchronizing entities. The Representation Protocol is centered on a few basic concepts. First, it introduces a consistent way to *identify* the data being synchronized—it provides flexible mechanisms to identify individual data items and sets of data items. Second, it provides a vocabulary to express various *operations* on data, such as insertion, modification, and deletion.

In this chapter we first review the above basic concepts. We then proceed further to discuss additional issues associated with any practical protocol. Such issues include protocol management, process flow, and text and binary representations. This chapter provides a comprehensive description of all aspects of the Representation Protocol with illustrative examples of individual Protocol elements.

Identifiers in SyncML

During synchronization, it is fundamentally important to consistently identify individual data items or a collection of data items. Sometimes it is necessary to identify the actual device or network server that is the intended recipient of a package. At other times it is necessary to identify a collection of data such as a database. At yet other times it may be necessary to identify individual data items in a datastore or a set of data items that satisfy certain criteria. SyncML uses the notion of `Target` and `Source` for identification. Each SyncML package contains `Target` and `Source` elements in various places. Depending on the context, the

`Target` and `Source` may refer to a particular machine, a particular datastore, a set of data items, or an individual data item.

Target and Source Addressing

Typically, each `Target` or `Source` element contains a Uniform Resource Identifier (URI) or a Uniform Resource Name (URN). The URI naming scheme is a flexible naming scheme that can be used to uniquely name any network resource, such as a machine, a datastore, or a data item. A URI begins with the identification of a scheme such as "http" or "IMEI" (International Mobile Equipment Identifier), and the part that follows the scheme is specific to the identified scheme. For example, "http://www.syncml.org/sync-Server" is a valid URI identifying a network Server in the "http" scheme. "IMEI:098712345" is a valid URI identifying a mobile device by its unique numeric identifier.

The meaning of the URI included in a `Target` or `Source` element inside a SyncML Message is completely determined by its context. For example, when used in the "header" part of a SyncML Message, the `Target` and `Source` refer to a mobile device or the network address of a Server. The header of a SyncML Message sent from a mobile Client to a Server may contain the following snippet:

```
<Target>
    <LocURI>http://www.syncml.org/sync-Server</LocURI>
</Target>
<Source>
    <LocURI>IMEI:098712345</LocURI>
</Source>
```

The `Target` and `Source` information in this case signify that the originating entity of the SyncML Message is a mobile device with a particular unique numeric identifier and the destination entity of the package is a network Server in the "syncml.org" domain.

In some cases, the `Target` or `Source` element may refer to the name of a datastore. For example, SyncML has some container elements, such as `Sync`, that refer to entire datastores and additionally contain other individual elements such as `Add` or `Replace`. The following snippet from a `Sync` command uses an absolute URI to refer to the datastore called "calendar" in the mobile device identified above.

```
<Source>
    <LocURI>IMEI:098712345/calendar</LocURI>
</Source>
```

Since the `Target` and `Source` elements are context sensitive, sometimes the specified URIs can be *relative* URIs. For example, within a SyncML Message where the header already uniquely identifies a device as the source, the following relative source address refers to the same absolute source address specified above.

```
<Source>
    <LocURI>./calendar</LocURI>
</Source>
```

During synchronization, identification of individual data items in a datastore is also required. For example, a `Replace` command must uniquely identify the actual data item being replaced. The following `Source` snippet extracted from a `Replace` command uses an absolute URI to signify that item "1001" from the calendar datastore of the mobile device is to be replaced.

```
<Source>
    <LocURI>IMEI:098712345/calendar/1001</LocURI>
</Source>
```

The context-sensitive `Target` and `Source` addressing in SyncML and the use of Uniform Resource Identifiers provide for a consistent way to identify machines, datastores, and individual data items. SyncML uses established standards, such as URIs or URNs, and attempts to provide a flexible, open identification scheme. In particular, SyncML does not enforce its own, specially designed identification scheme.

Target Address Filtering

Sometimes it is important to be able to address only a subset of a datastore. Consider, for example, a scenario where a mobile health care worker has lost all her appointment information while traveling. If she synchronizes with her Server application, an undesirable slow synchronization over a wireless link will be initiated, one that will involve the transfer of a large amount of appointment information—perhaps including appointments from the past few weeks and a few months into the future. At that moment, however, the mobile worker is most concerned with obtaining appointment information for that date only. Target address filtering will enable her to specify a selective target within a `Get` command that will fetch only the data that she needs most. The following snippet shows a Target Address Filter that can be used to have the desired effect of retrieving the calendar information for only January 5, 2002.

```
<Target>
    <LocURI>.calendar/Anna?DTSTART&GE;20020105T000000&AND;
    DTEND&LT;20020105T000000</LocURI>
</Target>
```

SyncML specifies search grammars that are to be used for various types of datastores, such as arbitrary databases (searched by unique identifiers), XML documents (searched by element type), contacts, calendars, and email.

Operations in SyncML

Users perform a variety of operations on data items. SyncML provides a set of *verbs*, or commands, to express different operations that are performed on data. The following discussion summarizes the various different operations, and their respective nuances and implications.

Modifying Data

Perhaps one of the most common operations that users perform is to modify a data item in a datastore. SyncML expresses this using the `Replace` element.

Different modifications have different semantics. Some modifications are only changes in fields within a data item, for example, changing the time field in a calendar data item. Some modifications may entail changing only certain meta-attributes of a data item, such as marking an email message from "unread" to "read."

Adding Data

The creation of new data entries is also a common operation. SyncML expresses this as the `Add` element. The `Add` operation is complicated because of the data identifier issue. SyncML allows Client devices and Servers to use different data identifiers. For example, new data items are added in a Server datastore. The Server only informs the Client that a set of items have been added and sends those items to the Client. As part of synchronization, the Client not only adds those items in the local datastore but also generates local unique identifiers for those items. The

Client then communicates those identifiers to the Server using a `Map` command.

Deleting Data

In addition to adding or modifying data items, users may wish to delete data items from a datastore. The `Delete` operation has various possible semantics.

The obvious kind of delete is when a user intends to delete a record from a datastore. This kind of delete is called a *hard delete*. During synchronization, the recipient of a hard delete operation, such as a Server, will actually delete the item from the corresponding Server datastore.

There is another notion of delete called the *soft delete*. The soft delete is inspired by the requirement that memory-constrained mobile devices may run out of space to store data and would like to be able to free up storage space by discarding certain data items, keeping only the most pertinent items. When a Server receives a soft delete operation, it does not delete the data item from its datastore; instead it will not send this item to the client again.

Another type of delete is the *archive delete*. Sometimes mobile Clients would like to delete a data item but want a Server to archive a copy of the data item. For example, an insurance worker may want to delete claims processing data from his mobile device for a claim that has just been completed, while wanting the Server datastore to archive the claim data.

Refreshing Data

It is not uncommon for a mobile device to have its data lost or corrupted by power failure, rough use, or other failures. It is desirable for mobile devices to be able to obtain the latest data from a Server. This can be thought of as a data refresh operation. The refresh operation is not required under normal circumstances, but SyncML recognizes the need for a practical protocol to support refresh semantics and supports it.

To initiate a Refresh From Server operation, the Client sends an `Alert` element to the Server with a "Refresh From Server" value. It is important to understand the distinction between refresh and slow synchronization. Slow synchronization entails transmission of all Client data items in a datastore to the Server, a comparison with all Server data items in the corresponding datastore, and resulting changes sent back to the Client. At the end of slow synchronization, the Client has a *synchronized* datastore, just as in ordinary synchronization. In contrast, the refresh operation only involves one-way data flow from the Server to the Client, obliterating the Client datastore with items from the Server datastore. At the end, the Client has a *copy* of the Server datastore. Symmetrically, SyncML also supports refreshing a Server datastore with values from a Client datastore.

Searching for Data

Mobile devices may require the ability to search for patterns within a data element or for particular elements within a datastore. SyncML has a Search command that enables the search of a datastore, with the resulting data returned in a Results command. A combination of the Get command with target address filtering, as discussed above, may also have the desired effect of a search operation. The Get operation can also be used to fetch complete datastores, such as device information. It is possible that the Put operation can be used instead of the Add or Replace operations, but the prevalent use of the Put operation entails replacing entire datastores, such as device information.

Grouped Operations on Data

All the typical user operations such as adding, replacing, and deleting data are individual operations on one or more data items. During synchronization, however, a datastore on a Client is synchronized with a datastore on a Server as a whole. It makes sense therefore to be able to group operations on a datastore within a *container* operation. The notion of a container simplifies a number of things. For example, authentication and access control could be performed at a datastore level for each container operation but not for individual operations inside. In an actual Server implementation, grouped operations in a container can be batched and scheduled efficiently. In addition, sometimes a group of operations have collective semantics in addition to individual operations inside.

The Sync container operation in SyncML allows multiple unit operations such as Add, Replace, and Delete to be grouped together in the context of one datastore. Aside from the grouping, and possible efficiencies derived from the grouping as indicated above, the Sync container has no additional semantics. The individual operations within a Sync container can be performed in any order. One or more operations within the Sync container can fail and not affect the status of the Sync. However, the Sync container can fail for other reasons, such as overall authentication failure with the datastore or inablility to communicate with the datastore.

Another possible container operation in SyncML is Atomic. Individual operations included within the Atomic container must all be successfully performed, or none at all will be. This container is useful when applications require transactional behavior. This container requires that the set of individual operations be performed but does not require them to be performed in the order specified. The Sequence container requires the individual operations that it contains to be performed in the order specified. This type of ordering is useful for applications that require workflow-like semantics. The Sequence container does not explicitly require that *all* operations be performed. The Sequence and the Atomic containers can be nested within each other to enforce strict ordering as well as transactional behavior.

Representation Protocol Elements

The messages exchanged during synchronization consist of a set of Elements. The format of these Elements is probably one of the more involved aspects of SyncML. Careful reading of the Representation Protocol specification will show that each Command Element has a clearly defined format and explicit list of allowed responses.

A more detailed look into the Representation Protocol reveals that it consists of several different types of Elements:

- The Message Container Elements
- The Protocol Management Elements
- The Command Elements
- The Common Use Elements
- The Data Description Elements

Proper combination of each of these Elements will create a well-formed SyncML Message. Note that not all Elements listed in the Representation Protocol are required to create a valid SyncML Message.

The Message Container Elements

The Message Container Elements are needed to encapsulate a SyncML Message. These three Elements are used to separate the header or session/message data from the command or synchronization data. The Message Container Elements are:

- SyncML

The container for a SyncML Message—i.e., it is the root element of the SyncML DTD. This only holds `SyncHdr` and `SyncBody`, so consider this similar to an envelope with the `SyncHdr` as the address and the `SyncBody` as the letter inside the envelope. Here is an example message without any real contents (only `SyncHdr` and `SyncBody`):

```
<SyncML xmlns='SYNCML:SYNCML1.0'>
    <SyncHdr>
    <!-- header information goes here -->
    </SyncHdr>
    <SyncBody>
    <!-- body information goes here -->
    </SyncBody>
</SyncML>
```

- SyncHdr

The `SyncHdr` contains the information about the message, similar to the address on an envelope. The `SyncHdr` holds information such as where the message is intended to go (`Target`), where the message is from (`Source`), the credentials of the sender, and which version of the Representation and Synchronization Protocols the message is using. Additionally, the `SyncHdr` can contain `Meta` information about the capabilities of the device, such as the maximum message size. If no response is desired to this message, a `NoResp` flag can be set in the `SyncHdr` as well.

Here is an example `SyncHdr`, coming from a Client:

```
<SyncHdr>
  <VerDTD>1.0</VerDTD>
  <VerProto>SyncML/1.0</VerProto>
  <SessionID>1</SessionID>
  <MsgID>1</MsgID>
  <Target>
    <LocURI>http://www.sync.org/servlet/syncit</LocURI>
  </Target>
  <Source>
    <LocURI>IMEI:001004FF1234567</LocURI>
  </Source>
</SyncHdr>
```

Note that this `SyncHdr` contains a `Source` that refers to IMEI. See the section in this chapter on `Source` and `Target` for more information on URIs.

- SyncBody

This is the body of the SyncML Message. The `SyncBody` contains a subset of the Command and Protocol Management Elements. The `SyncBody` is where the synchronization data and command information is stored; it is the content of the envelope. This is how the first `SyncBody` from a Client might look during an actual synchronization:

```
<SyncBody>
    <Alert>
      <CmdID>1</CmdID>
      <Data>200</Data> <!-- 200 = TWO_WAY_ALERT -->
      <Item>
    <Target><LocURI>./contacts/james_bond</LocURI></Target>
        <Source><LocURI>./dev-contacts</LocURI></Source>
        <Meta>
          <Anchor xmlns='syncml:metinf'>
            <Last>234</Last>
            <Next>276</Next>
          </Anchor>
        </Meta>
      </Item>
    </Alert>
</SyncBody>
```

This `SyncBody` contains an `Alert`, asking the Server to start a normal Two-way sync and indicating which datastores to synchronize. Note the use of change counters for the Sync Anchors.

The Protocol Management Elements

`Status` is the only Protocol Management Element:

- Status

`Status` is used to indicate the result of a command. This is not the same as `Results`, which returns `Search` information or `Get` information. There is one `Status` per command (for ALL commands), and they must be in the same order as the commands in the SyncML request. If needed, a `Status` can be sent against the `SyncHdr` by using a `CmdRef` of '0'. `Status` commands should be the first set of Elements in the `SyncBody`.

```
<Status>
    <CmdID>8</CmdID>
    <MsgRef>2<MsgRef>
    <CmdRef>9<CmdRef>
    <Cmd>Add</Cmd>
    <TargetRef>./bruce1</TargetRef>
    <SourceRef>IMEI:001004FF1234567</SourceRef>
    <Data>401</Data>
</Status>
```

This example of Status is for an Add command that failed due to lack of appropriate credentials. Note it is not possible to determine the datastore for the referred command without looking at the Status for the Sync command.

The Command Elements

In order to have any action happen in a session, a request must be sent. Typically, the Command Elements are those requests. These requests are a set of operations on one or more datastores.

The Command Elements are used to cause an action in the recipient of the SyncML message. A Command Element almost always contains data, either qualifying the command or conveying payload. The payload is something like a contact or event or email information. The Command Elements can be logically divided into three groups:

- Data Command Elements
- Datastore Command Elements
- Process Flow Command Elements

The Data Command Elements

The Data Command Elements are used to change application data. An example of application data might be contact information or a set of notes. The Data Command Elements are:

- Add

This command adds one or more payload items to a Datastore. If a Server sends an Add command, the implication is that the Client will send a Map command in return. However, if the Client uses the same UID as the Server, no Map will be required.

```
<Add>
    <CmdID>1<CmdID>
    <Item>
        <Source><LocURI>15</LocURI></Source>
```

```
        <Meta>
          <Type xmlns='syncml:metinf'>text/x-vcard</Type>
        </Meta>
        <Data>BEGIN:VCARD
          VERSION:3.0
          FN:Smith;Bruce
          N:Bruce Smith
          TEL;TYPE=WORK;VOICE:+1-919-555-1234
          END:VCARD
        </Data>
      </Item>
</Add>
```

This command adds a single business card in the vCard format. The Meta is used to indicate the media format of the payload. The Source indicates the UID of the item on the Server.

- Copy

This command creates a copy of an existing item, either in the same datastore or in another datastore. Copy can only create new items of the same type—it cannot make an event out of a contact, for example.

```
<Copy>
    <CmdID>12345</CmdID>
    <Item>
      <Target>
        <LocURI>mid:msg1@host.com</LocURI>
      </Target>
      <Source>
        <LocURI>./mail/bruce1/folders/Project%20XYZ</LocURI>
      </Source>
    </Item>
    <Item>
      <Target>
        <LocURI>mid:msg2@host.com</LocURI>
      </Target>
      <Source>
        <LocURI>./mail/bruce1/folders/Admin</LocURI>
      </Source>
    </Item>
</Copy>
```

This Copy command makes copies of two different email messages. This is done by specifying only the IDs of the email to copy (the Source) and the location to copy them to (the Target).

- Delete

This command removes an item from a datastore. It is also possible to use the soft delete option with this command, but that is rarely implemented. If soft delete is used on a recipient that does not support soft delete, then the delete can still be successful, but the delete will be a hard delete. More can be found on soft deletes in the sftDel section.

```
<Delete>
    <CmdID>13<CmdID>
    <Archive/>
    <Item>
      <Source><LocURI>15</LocURI></Source>
    </Item>
</Delete>
```

This command deletes a single item, but it also asks the recipient to archive the item prior to deletion. The Archive flag is an optional Element that is used to store items in something similar to a trash bin.

- Get

This command will retrieve an item, typically the Device Information. SyncML Data Synchronization only requires Get to return the Device Information; SyncML Device Management (see Chapter 9) will require support for more objects. It is possible to use Get to retrieve an object, but this usage is extremely rare and may not be supported.

```
<Get>
    <CmdID>3</CmdID>
      <Meta><Type xmlns='syncml:metinf'>
      application/vnd.syncml-devinf+xml</Type></Meta>
    <Item>
      <Target><LocURI>./devinf10</LocURI></Target>
    </Item>
</Get>
```

This example shows how to request the Device Information of the recipient. Note the use of the Meta to specify the type.

- Map

Map is used by Clients to tell a Server about a datastore that has had new items created by an Add command. The Map element contains one or more MapItems, as well as the Source and Target information.

```
<Map>
  <CmdID>13<CmdID>
  <Target>
      <LocURI>http://www.datasync.org/servlet/syncit</LocURI>
  </Target>
  <Source>
      <LocURI>IMEI:001004FF1234567</LocURI>
  </Source>
  <MapItem>
      <Target><LocURI>./0123456789ABCDEF</LocURI></Target>
      <Source><LocURI>15</LocURI></Source>
  </MapItem>
</Map>
```

This Map contains a single MapItem, but could contain more. However, each Map must deal with a single local and remote datastore.

- MapItem

The MapItem only holds two pieces of information: the Client UID in the Source, and the Server UID in the Target. The Client had received the Server UID earlier in an Add command. The Client is now telling the Server the mapping between the Server and Client UIDs. See Map for an example. The MapItem can also indicate a new UID for an existing item (e.g. a Client had to reissue UIDs to items). In that case, the new UID will be in the Source, and the old UID will be in the Target.

- Put

This will send an item, typically the Device Information. Note in SyncML Data Synchronization, the Put command is only required to put the Device Information.

```
<Put>
    <CmdID>2</CmdID>
    <Meta><Type xmlns='syncml:metinf'>
        application/vnd.syncml-devinf+xml</Type></Meta>
    <Item>
        <Source><LocURI>./devinf10</LocURI></Source>
        <Data>
            <DevInf xmlns='syncml:devinf'>
              <Man>Big Factory, Ltd.</Man>
              <Mod>4119</Mod>
              <OEM>Jane's phones</OEM>
              <DevId>1218182THD000001-2</DevId>
              <DevTyp>phone</DevTyp>
            </DevInf>
        </Data>
    </Item>
</Put>
```

This example of the Put shows the Device Information (see Chapter 7) being sent to the recipient. Note this is a very truncated version of the Device Information.

- Replace

The Replace command is used to update an existing item with new information. The Replace command contains one or more payload items, each with their own UID. Here is an example Replace command:

```
<Replace>
    <CmdID>4</CmdID>
    <Meta>
        <Type xmlns='syncml:metinf'>text/x-vcard</Type>
    </Meta>
    <Item>
        <Source><LocURI>1012</LocURI></Source>
        <Data>BEGIN:VCARD
            VERSION:3.0
            FN:Mueller;Bruce
            N:Bruce Mueller
            TEL;TYPE=WORK;VOICE:+49 (7031) 16-5509
            END:VCARD
        </Data>
    </Item>
</Replace>
```

The Datastore Command Elements

The Datastore Command Elements are used to define actions pertaining to an entire Datastore. The Datastore Command Elements are:

- Alert

Alert is used to signal special "nonstandard" commands, such as notifications, synchronization requests, and text displays. The most typical use for Alert is to ask for a specific type of synchronization, such as a normal Two-way sync or a Slow sync.

```
<Alert>
    <CmdID>509</CmdID>
    <Data>200</Data>
    <Item>
        <Target><LocURI>./contacts/james_bond</LocURI></Target>
        <Source><LocURI>./dev-contacts</LocURI></Source>
        <Meta>
            <Anchor xmlns='syncml:metinf'>
                <Last>234</Last>
                <Next>243</Next>
            </Anchor>
```

```
        </Meta>
    </Item>
</Alert>
```

This `Alert` is requesting a normal Two-way synchronization. It is also specifying the sync anchors, as a means of allowing the recipient to verify that a normal sync is acceptable. If the recipient does not agree with the sync anchors, the recipient will ask for a Slow sync instead.

- Results

`Results` returns the output from a `Search` command or a `Get` command. `Results` stores the output in (one or more) `Item`. `SourceRef` and `TargetRef` are the `Source` and `Target` from the command with which `Results` is associated.

```
<Results>
    <CmdID>42<CmdID>
    <Meta><Type xmlns='syncml:metinf'>
             application/vnd.syncml-devinf+xml</Type></Meta>
    <Item>
        <Source><LocURI>devinf10</LocURI></Source>
        <Data>
             <!-- devinfo data would go here -->
        </Data>
    </Item>
</Results>
```

`Results` is typically used to return the Device Information. The example above shows how `Results` would return the Device Information, but the actual data is excluded from the example to keep it short.

- Search

`Search` asks for a search on a Datastore. If the `Search` command contains the `NoResults` flag, then the output from the search must be placed in temporary storage, as specified in the `Target`. It is necessary to be aware that `Search` can have more than one `Source`, meaning the `Search` command must search more than one datastore at once. This can be a powerful command, one that can easily overwhelm a product not prepared for a large `Results` message. Note the search grammar must be specified in the `Meta`, and a nonsupported search grammar will be rejected. At this time, there is no well-defined search grammar in use.

```
<Search>
    <CmdID>1234</CmdID>
    <Cred>
        <Meta>
            <Type xmlns='syncml:metinf'>syncml:auth-md5</Type>
            <Format xmlns='syncml:metinf'>b64</Format>
        </Meta>
        <Data>OGNkNDI1ZTZjNjgwMTNiYWZkOWEyN2JjMjN1ZDM4YzENCg==</Data>
    </Cred>
    <Source>
    <LocURI>http://www.sync.org/servlet/syncit/bruce1</LocURI>
    </Source>
    <Meta><Type xmlns='syncml:metinf'>application/sql</Type></Meta>
    <Data>SELECT * WHERE "FN" EQ "Bruce Smith"</Data>
</Search>
```

This search is slightly different from normal, as it uses SQL to locate a contact that contains the name Bruce Smith. A more typical search would use a CGI script similar to the Target Address Filtering. It also contains credentials in MD5 format, as the datastore being queried requires authentication.

- Sync

Sync is the command used to perform the synchronization between two datastores. It holds one or more Data Commands (e.g. Add, Delete, and Replace), specifies the local (Source) and remote (Target) datastores, specifies any credentials necessary for synchronization with the remote datastore, and can indicate the media format for the data commands. Sync must be used whenever two datastores need to be synchronized. If you just wish to store an object within another datastore, and do not wish to synchronize the entire datastore, then just use the Put command.

```
<Sync>
    <CmdID>3</CmdID>
    <Target><LocURI>./contacts/james_bond</LocURI></Target>
    <Source><LocURI>./dev-contacts</LocURI></Source>
    <Meta>
        <Mem xmlns='syncml:metinf'>
            <FreeMem>8100</FreeMem>
            <!--Free memory (bytes) in Calendar database -->
            <FreeId>41</FreeId>
            <!--Number of free records in Calendar database-->
        </Mem>
    </Meta>
    <Replace>
        <CmdID>4</CmdID>
```

```
            <Meta>
                <Type xmlns='syncml:metinf'>text/x-vcard</Type>
            </Meta>
            <Item>
                <Source><LocURI>1012</LocURI></Source>
                <Data><!--The vCard data would be placed here.--></Data>
            </Item>
        </Replace>
</Sync>
```

This Sync example includes Mem—information about the available memory and/or IDs. This particular example has 8100 bytes free, and 41 free IDs. The Replace that is in the Sync only has one Item, which is a vCard. The Mem hint is usually in the Sync command to help the recipient determine whether or not their new data (to be sent back) will fit into the datastore.

The Process Flow Command Elements

The Process Flow Command Elements are used to control the processing of the various Data Command and Datastore Command Elements. In SyncML Data Synchronization, these Command Elements are currently optional and, as a result, are not implemented very often. However, in SyncML Device Management, they are mandatory. The Process Flow Command Elements are:

- Atomic

Atomic requires that all subcommands be executed, or none will be executed. This requires a rollback capability, similar to Database Transaction Processing. This is optional for SyncML Data Synchronization (but not for SyncML Device Management) and tends to be difficult to implement properly. As a result, it is not implemented very often. This command may be useful for applications synchronizing Relational Databases that require transactional guarantees, such as a sales application.

```
<Atomic>
    <CmdID>321</CmdID>
    <Add>
        <CmdID>322</CmdID>
        <Item>
            <Target><LocURI>./devinf10/pen</LocURI></Target>
            <Data>Yes</Data>
        </Item>
    </Add>
    <Replace>
        <CmdID>323</CmdID>
```

```
            <Item>
              <Target><LocURI>./devinf10/version</LocURI></Target>
                <Data>20020401T133000Z</Data>
            </Item>
        </Replace>
</Atomic>
```

This `Atomic` sequence is a Device Management sequence. The "pen" setting is being set to "Yes", and the version is being set to a date in 2002. Note that if `Add` had failed, then the `Replace` would not execute, and the `Atomic` would fail. Also note that if the `Replace` failed, then the `Add` would be unexecuted (or rolled back in Database terms), and the `Atomic` would fail.

- Sequence

`Sequence` requires all subcommand Elements to be executed in the sequence they appear. Again, this is optional in SyncML Data Synchronization, but required in SyncML Device Management.

```
<Sequence>
    <CmdID>1234</CmdID>
    <Add>
        <CmdID>1235</CmdID>
        ...
    </Add>
    <Add>
        <CmdID>1236</CmdID>
  ...
    </Add>
    <Delete>
        <CmdID>1237</CmdID>
  ...
    </Delete>
</Sequence>
```

If the recipient of this `Sequence` was unable to guarantee that each command was executed in order, then all of those commands would fail, and the `Sequence` command would fail. Otherwise, the recipient will execute the commands in order, and return the appropriate `Status` from each command.

Note that `Atomic` does not imply that each command must be done in sequence, nor does `Sequence` imply that all of the subcommands must succeed. These are complimentary commands.

The Common Use Elements

The Common Use Elements are so named because they can appear as subelements in most of the Command and SyncHdr Elements. They are used to provide a set of common functions and to help reduce the amount of special parsing code. Imagine the size of a SyncML product if it had to accommodate all these common elements if they were not specifically called out as common code elements—the amount of redundant code would be staggering. The Common Use Elements are:

- Chal

Chal (short for Challenge) is used for authentication purposes. Chal allows products to ask for credentials in a specific format. For example, if a Client produces credentials in the Basic format and the Server prefers credentials in the MD5 format, the Server can request that the Client send the credentials in the MD5 format via the Chal Element.

```
<Status>
    <MsgRef>1</MsgRef>
    <CmdRef>0</CmdRef>
    <Cmd>SyncHdr</Cmd>
    <TargetRef>http://www.datasync.org/servlet/syncit</TargetRef>
    <SourceRef>IMEI:001004FF1234567</SourceRef>
    <Chal>
        <Meta>
            <Type xmlns='syncml:metinf'>syncml:auth-basic</Type>
            <Format xmlns='syncml:metinf'>b64</Format>
        </Meta>
    </Chal>
    <Data>401</Data>
</Status>
```

This Status command contains a Chal asking for the credentials in Basic format.

- Cmd

Cmd contains the ASCII name of the command (useful for debugging).

```
<Status>
    <MsgRef>1</MsgRef>
    <CmdRef>14</CmdRef>
    <Cmd>Add</Cmd>
    <TargetRef>./mail/bruce1</TargetRef>
    <Data>401</Data>
</Status>
```

This Status shows that the command was an Add command.

- CmdID

CmdID contains the ID of the command within the SyncML message. CmdID is reference in Status. A good method is to use an unsigned integer (16 bits are sufficient) for this value, starting with '1' and then incrementing the counter after each use. Note that a CmdID ('0') is reserved to indicate the SyncHdr.

```
<Delete>
    <CmdID>3456</CmdID>
    <Item>
        <Target><LocURI>./11</LocURI></Target>
    </Item>
</Delete>
```

This Delete command has a CmdID of 3456, a value that will rarely be seen, assuming a start value of '1' and incrementing by one.

- CmdRef

CmdRef is the ID of the command being referred to in Status.

```
<Status>
    <CmdID>1</CmdID>
    <MsgRef>2</MsgRef>
    <CmdRef>0</CmdRef>
    <Cmd>SyncHdr</Cmd>
    <TargetRef>http://www.syncml.org/sync-Server</TargetRef>
    <SourceRef>IMEI:493005100592800</SourceRef>
    <Data>101</Data> <!--Statuscode for Busy-->
</Status>
```

Note in this example, the CmdRef is '0'; this explicitly refers to the SyncHdr. This Status is telling the caller that the device is busy.

- Cred

Cred contains a set of credentials in a particular format. SyncML only requires support for either Basic or MD5. Additional authentication standards can be used, such as X.509.

```
<Cred>
    <Meta>
        <Type xmlns='syncml:metinf'>syncml:auth-basic</Type>
    </Meta>
    <Data>QnJ1Y2UyOk9oQmVoYXZ1</Data>
    <!-- base64 formatting of "userid:password" -->
</Cred>
```

- Final

Final is a flag indicating the end of a package. It is helpful to think of Final as a baton that is passed between the Client and the Server. Since the Client starts the session, the Client holds the Final "baton" and passes it to the Server when done with its package. Likewise, the Server will pass the Final "baton" back when done with its package. This baton passing will continue until the session is done.

```
<SyncML xmlns='SYNCML:SYNCML1.0'>
    <SyncHdr>...</SyncHdr>
    <SyncBody>
        ...
        <Final/>
    </SyncBody>
</SyncML>
```

This is an abbreviated message, showing only the location of the Final flag. Final must be the last command in the SyncBody.

- Lang

Lang allows the product to request a particular language for a payload.

```
<Get>
    <CmdID>12</CmdID>
    <Lang>en-US</Lang>
    <Item>
        <Target><LocURI>./telecom/pb</LocURI></Target>
        <Source>
            <LocURI>http://www.sync.com/servlet/</LocURI>
        </Source>
        <Meta>
            <Type xmlns='syncml:metinf'>text/x-vCard</Type>
        </Meta>
    </Item>
</Get>
```

This Get command asks for a vCard to be delivered with US English language.

- LocName

LocName is the human readable display name for a target or address. Note when MD5 is used, LocName contains the username as an aid in verifying the MD5 digest (since the MD5 digest is a one-way hashing algorithm).

```
<SyncHdr>
    <VerDTD>1.0</VerDTD>
    <VerProto>SyncML/1.0</VerProto>
    <SessionID>1</SessionID>
    <MsgID>1</MsgID>
    <Target>
        <LocURI>http://www.syncml.org/sync-Server</LocURI>
    </Target>
    <Source>
        <LocURI>IMEI:493005100592800</LocURI>
        <LocName>Bruce2</LocName> <!-- userId -->
    </Source>
    <Cred>
        <Meta><Type xmlns='syncml:metinf'>
            syncml:auth-md5</Type>
        </Meta>
        <Data>NTI2OTJhMDAwNjYxODkwYmQ3NWUxN2RhN2ZmYmJlMzk=</Data>
        <!-- Base64 coded MD5 digest of "Bruce2:OhBehave:Nonce" -->
    </Cred>
</SyncHdr>
```

This SyncHdr contains credentials in MD5 form, and to aid in the verification of these credentials, the Source contains LocName with the username.

- LocURI

LocURI contains a specific target or address. LocURI is used in several commands and can point to either the Target or Source. In any case, it is important to make sure whether the address should be referred to as an absolute or a relative URI. Absolute URIs are easy to spot, as they typically start with something like "http://" or "ftp://". Relative URIs are more like "./contact/james_bond" or "event/bruce". Note in relative URIs, "./bruce" and "bruce" are the same thing. Usage of relative URIs is encouraged whenever possible, as this will reduce the number of bytes transferred–an important aspect to remember when dealing with wireless devices (which can cost on a per-byte basis).

```
<SyncHdr>
    <VerDTD>1.0</VerDTD>
    <VerProto>1.0</VerProto>
    <SessionID>1</SessionID>
    <MsgID>1</MsgID>
    <Target><LocURI>http://www.syncml.host.com/</LocURI></Target>
    <Source><LocURI>IMEI:001004FF1234567</LocURI></Source>
</SyncHdr>
```

This shows LocURI being used in both the Target and the Source.

- MoreData

Added in version 1.1, `MoreData` is used to signal to the recipient that the data being sent in the parent command is incomplete–the rest of the data will come in later message(s). `MoreData` cannot be used if the data can fit into a message.

```
<Add>
    <CmdID>15</CmdID>
        <Meta>
        <Type>text/x-vcard</Type>
        <Size>3000</Size>
    </Meta>
    <Item>
        <Source><LocURI>2</LocURI></Source>
        <Data>BEGIN:VCARD
            VERSION:2.1
            FN:Bruce Smith
            N:Smith;Bruce
            TEL;WORK;VOICE:+1-919-555-1234
            TEL;WORK;FAX:+1-919-555-9876
            NOTE: here starts a huge note field, or icon etc...
        </Data>
        <MoreData/>
    </Item>
</Add>
```

The `Add` command contains a single `Item`. The `Item` has a single large piece of payload data. The `Item` element also has the `MoreData` element, indicating the data is not complete. The recipient would accept this `Add` by sending back a `Status` with "Chunked item accepted and buffered".

- MsgID

`MsgID` is the ID of the current message–useful for later reference to a command. `MsgID` only exists in the `SyncHdr`.

```
<SyncHdr>
    <VerDTD>1.0</VerDTD>
    <VerProto>SyncML/1.0</VerProto>
    <SessionID>1</SessionID>
    <MsgID>1</MsgID>
    <Target><LocURI>IMEI:493005100592800</LocURI></Target>
    <Source>
        <LocURI>http://www.syncml.org/sync-Server</LocURI>
    </Source>
</SyncHdr>
```

The MsgID for this SyncHdr is '1', meaning it is the first message from this device in this session.

- MsgRef

MsgRef is the ID of a message containing a command that is being referred to (typically in Status, but it can also appear in Results). This is useful if you are trying to respond to a backlog of commands. It is also useful for a sanity check in the recipient.

```
<Status>
    <CmdID>1</CmdID>
    <MsgRef>1</MsgRef>
    <CmdRef>0</CmdRef>
    <Cmd>SyncHdr</Cmd>
    <TargetRef>http://www.syncml.org/sync-Server</TargetRef>
    <SourceRef>IMEI:493005100592800</SourceRef>
    <Chal>
        <Meta>
            <Type xmlns='syncml:metinf'>syncml:auth-basic</Type>
            <Format xmlns='syncml:metinf'>b64</Format>
        </Meta>
    </Chal>
    <Data>407</Data> <!--Credentials missing-->
</Status>
```

The Status command is referring to the SyncHdr and is indicating that the necessary credentials were missing. The MsgRef in this Status is '1', indicating that it is the first message of the session.

- NoResp

NoResp is a flag indicating that no response is wanted or expected for a command. This flag may also be set in the SyncHdr if no response is wanted at all in return.

```
<SyncHdr>
    <VerDTD>1.0</Version>
    <VerProto>SyncML/1.0</VerProto>
    <SessionID>1</SessionID>
    <MsgID>2</MsgID>
    <Target><LocURI>IMEI:001004FF1234567</LocURI></Target>
    <Source>
        <LocURI>http://www.sync.org/servlet/syncit</LocURI>
    </Source>
    <NoResp/>
</SyncHdr>
```

This Message is being built with the expectation that no response is needed or wanted. The SyncHdr contains the NoResp flag, and for no responses to come back, any other commands, such as Replace or Sync, will also need the NoResp flag.

- NoResults

NoResults is a flag indicating that the results from a Get or Search must be held in temporary storage as a source in a later command. It is possible that the recipient of this command may return an error if it cannot create temporary storage. An implementation should be careful about using this flag, as some recipients cannot support it.

```
<Search>
    <CmdID>3</CmdID>
    <NoResults/>
    <Source><LocURI>./bruce1/emp_tabl.db</LocURI></Source>
    <Target><LocURI>./bruces</LocURI></Target>
    <Meta><Type xmlns='syncml:metinf'>application/sql</Type></Meta>
    <Data>SELECT * WHERE "FN" EQ "Bruce Smith"</Data>
</Search>
```

The Search command contains the flag NoResults, indicating that the results of the search must be stored in the temporary location "./bruces".

- NumberOfChanges

Added in version 1.1, NumberOfChanges can only be used in the Sync command. NumberOfChanges is used to inform the recipient how many items will be sent from a particular datastore. The recipient may use this data to show the synchronization progress.

```
<Sync>
    <CmdID>5</CmdID>
    <Target><LocURI>contacts</LocURI></Target>
    <Source><LocURI>Contact/Unfiled</LocURI></Source>
    <NumberOfChanges>23</NumberOfChanges>
    ...
</Sync>
```

This example Sync indicates that 23 changes will be coming from this particular datastore.

- RespURI

RespURI is a URI that must be used for the response to the current message. This command allows for direction of synchronization messages to different Servers in a Server farm. This command can also be used to pass session data to a Client, and the Client must pass this information back in the form of the Target URI.

```
<SyncHdr>
    ...
    <RespURI>http://www2.datasync.org/servlet/syncit/bruce1?sessioninfo
            =01456ac2354ace</RespURI>
</SyncHdr>
```

RespURI can also be used to have the Client pass back session information. In this example, the Client is being told to respond to www2.datasync.org and, in addition, pass along some session information strings. This is a useful property when dealing with Server farms, or when one wants to have session information up front. RespURI is also helpful in that the Client sets the SessionID, and the Server cannot change that, so it has to use the RespURI instead.

- SessionID

SessionID is the ID of the current session. This is useful for determining if the message is from the current session, or if the session has restarted. The SessionID can remain the same value indefinitely, but this kind of usage greatly increases the difficulty in debugging your product and is highly discouraged.

```
<SyncHdr>
    <VerDTD>1.0</VerDTD>
    <VerProto>SyncML/1.0</VerProto>
    <SessionID>42</SessionID>
    <MsgID>1</MsgID>
    ...
</SyncHdr>
```

This would be the SyncHdr from the first message in a session. The SessionID is 42 in this case.

- SftDel

SftDel is a flag indicating that a delete is a soft delete. Typically, this command will come from the Server as a means of clearing extra space on the Client. The Client must maintain the LUID of the soft-deleted item, in case a Server wants to update that object at a later date. A Client may also send this flag, indicating to the Server that it does not want this object again, but not to delete it.

```
<Delete>
    <CmdID>4</CmdID>
    <SftDel/>
    <Item>
       <Target><LocURI>./11</LocURI></Target>
    </Item>
</Delete>
```

        This Delete command specifies that a soft delete needs to be done on the item with ID of '11'.

- Source

        Source specifies the source routing or mapping information. Source exists in many commands and indicates many different things, but it always indicates local data or routing. For example, in the Map command, it indicates the local routing information for the datastore that generated mapping.

```
<Map>
    <CmdID>5</CmdID>
    <Target><LocURI>./contacts/sspielberg</LocURI></Target>
    <Source><LocURI>./tables</LocURI></Source>
    <MapItem>
        <Target><LocURI>./0123456789ABCDEF</LocURI></Target>
        <Source><LocURI>./12</LocURI></Source>
    </MapItem>
</Map>
```

        This Map has Source used in two places: the Map and the MapItem. Source in Map refers to the datastore, and Source in the MapItem refers to the item in the datastore.

- SourceRef

        When present, SourceRef contains the value that was in the Source in the referenced command. When used in a Status command, it specifies the source address (LocURI) of the object. When used in the Results command, it contains the associated datastore URIs.

```
<Status>
    <CmdID>2</CmdID>
    <MsgRef>1</MsgRef>
    <CmdRef>5</CmdRef>
    <Cmd>Replace</Cmd>
    <SourceRef>1212</SourceRef>
    <Data>208</Data> <!-- Conflict, originator wins -->
</Status>
```

The Status command shown here uses the SourceRef to contain the value that was in the Source in the Replace command.

- Target

Target specifies the target routing or mapping information. Target is similar to Source, but instead refers to the destination. For example, Target in SyncHdr refers to the routing information for the network device that is receiving the message.

```
<Sync>
    <CmdID>3</CmdID>
    <Target><LocURI>./calendar/james_bond</LocURI></Target>
    <Source><LocURI>./dev-calendar</LocURI></Source>
    ...
</Sync>
```

Target in Sync refers to the datastore, and in this case, the Sync wants to operate on a secret agent's calendar.

- TargetRef

When present, TargetRef contains the value that was in the Target in the referenced command. When used in a Status command, it specifies the target address (LocURI) of the object. When used in the Results command, it contains the temporary datastore specified in the Search command.

```
<Status>
    <CmdID>1</CmdID>
    <MsgRef>1</MsgRef>
    <CmdRef>0</CmdRef>
    <Cmd>SyncHdr</Cmd>
    <TargetRef>http://www.syncml.org/sync-Server</TargetRef>
    <SourceRef>IMEI:493005100592800</SourceRef>
    <Data>212</Data> <!--Authenticated for session-->
</Status>
```

This is a Status for a SyncHdr that indicates the credentials were accepted and need not be sent for the rest of the session. In this Status, the TargetRef contains the value of the Target in the previous message.

- VerDTD

VerDTD specifies the major and minor version of the SyncML Representation Protocol. This is not the same as the Synchronization Protocol, though. This is an important field to pay attention to, as different major versions are not required to be compatible. Devices supporting 1.1 should note that if they receive a message with a VerDTD of 1.0, then they should send a VerDTD of 1.0 (and use the 1.0 DTD), as that sender can only handle the version it specified. The VerDTD can be seen in this SyncHdr.

```
<SyncHdr>
    <VerDTD>1.0</VerDTD>
    <VerProto>SyncML/1.0</VerProto>
    ...
</SyncHdr>
```

- VerProto

VerProto specifies the protocol used, as well as the major and minor version of the Synchronization Protocol. Again, it is important to pay attention to this, since the protocol could be SyncML or SyncML DM. See VerDTD for an example.

The Data Description Elements

The Data Description Elements are:

- Data

Data contains discrete SyncML data or payload; this is where the actual data is stored. Examples are credentials, contact information, and status code.

```
<Status>
    <CmdID>1</CmdID>
    <MsgRef>1</MsgRef><CmdRef>0</CmdRef><Cmd>SyncHdr</Cmd>
    <TargetRef>http://www.syncml.org/sync-Server</TargetRef>
    <SourceRef>IMEI:493005100592800</SourceRef>
    <Data>212</Data>
    <!--Statuscode for OK, authenticated for session-->
</Status>
```

This Status command contains Data indicating that the credentials are valid for the entire session. This works well unless there are large communication delays.

- Item

Item is a container for data; it holds information about the local data (Source), information about the remote data (Target), the data itself, and any Meta information on the data. Item is used to separate the data and its identification from the command information.

This example Delete command contains two Items:

```
<Delete>
    <CmdID>6</CmdID>
    <Item>
        <Target><LocURI>./11</LocURI></Target>
    </Item>
    <Item>
        <Target><LocURI>./12</LocURI></Target>
    </Item>
</Delete>
```

This example Add command contains a vCard.

```
<Add>
    <CmdID>1234</CmdID>
    <Item>
        <Source><LocURI>12</LocURI></Source>
        <Meta>
            <Type xmlns='syncml:metinf'>text/x-vCard</Type>
        </Meta>
        <Data>BEGIN:VCARD
VERSION:2.1
FN:Smith;Bruce
N:Bruce Smith
TEL;TYPE=WORK;VOICE:+1-919-555-1234
END:VCARD
        </Data>
    </Item>
</Add>
```

- Meta

    Meta is used to specify meta-information about the parent Element. For example, Meta can be used to specify the format of the data or to indicate the size of an object.

```
<Cred>
    <Meta>
        <Type xmlns='syncml:metinf'>syncml:auth-md5</Type>
        <Format xmlns='syncml:metinf'>b64</Format>
    </Meta>
    <Data>OGNkNDI1ZTZjNjgwMTNiYWZkOWEyN2JjMjNlZDM4YzENCg==</Data>
</Cred>
```

This Cred contains a Meta; the Meta contains a Type and a Format command. These Meta commands are used in processing the credentials.

Text and Binary Representation

All the examples in this chapter so far have used text encoding, or XML encoding. It is worth noting that SyncML also provides binary encoding—WXBML [WBXML01]. WBXML is the preferred method of encoding when operating over limited bandwidth media. Conversion from XML to WBXML is done by replacing the verbose tags with binary tokens. These binary tokens can have a value from 1 to 127 (0x00 to 0x8f). With the limitation of only 127 tags, it is necessary to sometimes use a different "code page" to continue the token representation for a common DTD. In fact, the first possible token is the flag for changing code pages. The SyncML Representation binary tokens can be found in section 8 of the Representation Document.

Note when using the Meta information or the Device Information, a code page switch is necessary. The tags for the MetInfo page are defined in the Meta Information Specification. The tags for the Device Information space are defined in the Device Information Specification. Also note that the Device Information does not use the same code space as the SyncML Representation and Meta Information.

Examples of WBXML may be found in the SyncML Synchronization Protocol document, as well as in Chapter 7.

Static Conformance Requirements

The Representation Protocol also defines a set of Static Conformance Requirements (SCR) tables. These tables define what a product is required to support. The tables also indicate optional commands or Elements. Here is an example from the SCR tables:

Command	Support of Synchronization Server		Support of Synchronization Client	
	Sending	Receiving	Sending	Receiving
Chal	MUST	MUST	MAY	MUST

This example is from the Common Use Elements section of the SCR tables. For a Server, the Chal command is required for sending and receiving. However, for a Client, it is only required to receive the Chal command.

According to the RFC 2119, MAY indicates entirely optional, SHOULD indicates strongly suggested (i.e. a good technical reason for not implementing), and MUST indicates required.

7

Supportive SyncML Components

The Representation Protocol [SRP02] and the Synchronization Protocol [SSP02] are the main components of SyncML® Data Synchronization technology. A few important elements still need to be added to the main protocols to realize the complete SyncML framework. These make the SyncML framework comprehensive such that it can efficiently be used for its designed functionality.

To complete the SyncML framework, there are basically two types of elements still to be considered. First, the SyncML protocols need some supportive elements. To address this, the Meta Information (MetInf) DTD [SMI02] and the Device Information (DevInf) DTD [SDI02] are introduced in this framework.

A second missing link is related to connectivity, or how SyncML Messages are transferred from one entity to another. Transport protocols of different kinds are required. The SyncML framework provides this transport functionality in the HTTP [SHB02], WSP [SWB02], and OBEX [SOB02] bindings.

These missing elements have important roles in finalizing SyncML technology. The additional components complete the picture and show how different SyncML components work with each other.

SyncML Architecture and Components

As described in Chapter 5, the Synchronization Protocol uses the services of the Representation Protocol, the MetInf DTD, and the DevInf DTD to create meaningful SyncML Messages and sessions. They are an

Chapter 7 ▸ Supportive SyncML Components

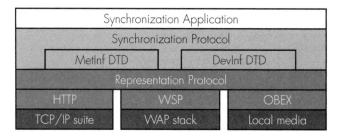

Figure 7–1
SyncML Protocol architecture

integral part of the SyncML technology. Their position in the SyncML Data Synchronization Protocol stack is depicted in Figure 7–1. The element types of the MetInf DTD and the DevInf DTD are used inside the SyncML DTD as defined by the Representation Protocol. Consequently, the Synchronization Protocol utilizes the services of all three of these components.

Products enabling SyncML are required to support the MetInf and DevInf DTDs. With the transport protocols, there are more choices depending on the environment at which a product is targeted. The specified SyncML Transport Bindings for HTTP, WSP, and OBEX offer a comprehensive set of protocols to be utilized. For instance if a synchronization server product is designed to offer a network service, HTTP is a natural choice. Similarly, the WSP binding is the usual selection for a mobile device that supports WAP.

Figure 7–2 depicts an example of a SyncML Message sent from a Client to a Server. The example also shows how the Meta Information DTD and the Device Information DTD are utilized inside a SyncML Message. In addition, the Message shown is further encapsulated in a HTTP message.[1] Further, various element types of the Representation Protocol encapsulate the element types of the MetInf and DevInf DTDs.

Figure 7–3 depicts a SyncML session as defined by the Synchronization Protocol. In addition, it shows how HTTP is used to carry SyncML Messages back and forth, and how a transport connection is enabled between a SyncML Client and Server. In the figure, the SyncML Client acts as a HTTP client and the SyncML Server as a HTTP server.

---

1. Note that the example is for illustrative purposes and may not exactly reflect the specifications. Parts of the mandatory HTTP headers and SyncML commands are removed from the example to keep it short and understandable.

SyncML Architecture and Components

Headers of transport protocol
```
POST ./servlet/syncit HTTP/1.1
Host: http://syncml-server.org
Accept-Charset: utf-8
Accept-Encodings: chunked
Content-Type: application/vnd.syncml+xml; charset="utf-8"
Transfer-Encoding: chunked
```

```
<SyncML>
<SyncHdr>
  <VerDTD>1.0</VerDTD><VerProto>SyncML/1.0</VerProto>
  <SessionID>1</SessionID><MsgID>1</MsgID>
  <Target><LocURI>http://syncml-server.org</LocURI></Target>
  <Source><LocURI>IMEI:4930051005928</LocURI></Source>
</SyncHdr>
<SyncBody>
  <Put>
    <CmdID>2</CmdID>
```

Usage of Meta Information
```
    <Meta>
      <Type xmlns='syncml:metinf'>
      application/vnd.syncml-devinf+xml
      </Type>
    </Meta>
```

```
    <Item>
      <Source><LocURI>./devinf10</LocURI></Source>
```

Usage of Device Information
```
      <Data>
        <DevInf xmlns='syncml:devinf'>
          <Man>Big Factory, Ltd.</Man>
          <Mod>4119</Mod>
          <OEM>Jane's phones</OEM>
        </DevInf>
      </Data>
```

```
    </Item>
  </Put>
</SyncBody>
</SyncML>
```

Figure 7–2
An example of a SyncML Message within an HTTP message

A SyncML Message formatted according to the Representation Protocol is only encapsulated by a message of the transport protocol and then transferred between a Client and a Server.

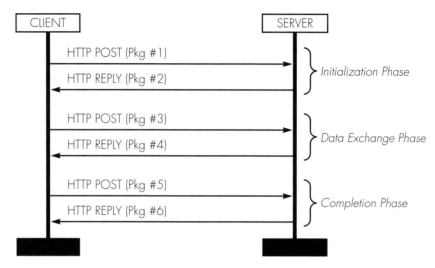

Figure 7-3
SyncML session and usage of HTTP

Complementary DTD Components

The complementary DTDs of SyncML are the Meta Information DTD and the Device Information DTD. They are used by the protocols based on the usage of the Representation Protocol: the Synchronization Protocol and the Device Management Protocol [SDP02]. The Synchronization Protocol utilizes both the Meta Information DTD and the Device Information DTD. The Device Management Protocol only uses the Meta Information DTD.

Meta Information DTD

The purpose of the Meta Information DTD (MetInf DTD) is to provide a set of markups to be used by the SyncML DTD (i.e. the Representation Protocol) to identify meta-information associated with a SyncML command or a data item. In practice, it is hard to construct fully operational and meaningful SyncML Messages without the MetInf DTD (e.g., the format or type of a data item).

There are a couple of clear advantages to separating the MetInf DTD from the Representation Protocol. First, the MetInf DTD can be reused with other technologies in addition to SyncML. Second, the representation of meta-information within the Representation Protocol is not tied to specific element types. This allows for additional meta-information element types targeted for specific purposes to be easily utilized within the Representation Protocol.

The Meta Information DTD has a simple, flat structure. There are only two element types (Anchor and Mem) that have child elements. Actually, in practice, the usage of MetInf is even simpler. When the MetInf DTD is used within the Representation Protocol, the root element type (MetInf) is not used at all. The XML example below shows how to embed an element of the Meta Information DTD (NextNonce) inside a command of the Representation Protocol.

```
<Status>
  <CmdID>12</CmdID><MsgRef>3</MsgRef>
  <CmdRef>45</CmdRef><Cmd>Add</Cmd>
  <TargetRef>./sync_DB</TargetRef>
  <Data>401</Data>
  <Item>
    <Meta>
      <NextNonce xmlns='syncml:metinf'>
      ZG9iZWhhdmUNCg==
      </NextNonce>
    </Meta>
  </Item>
</Status>
```

Although the MetInf DTD could be used outside SyncML, there are characteristics that closely link it to SyncML. For instance, the use of many element types of the MetInf DTD is not compelling outside a synchronization framework like SyncML. In addition, the MetInf DTD uses the same WBXML (Wireless Binary XML) binary tag code space[2] as the Representation Protocol, although the MetInf element types belong to a different XML namespace.

MetInf element types

The element types of the MetInf DTD can be divided into three different categories. The element types can be either related to actual content transferred in a SyncML Message, or to dynamic device characteristics, or to miscellaneous purposes. Table 7-1 shows the categories and the element types that belong to them.

---

2. The WBXML binary tag code space defines binary tags to corresponding clear-text XML elements types.

Table 7-1
MetInf Element Types within Categories

Category	Element Types
Content related	Format, Mark, Size, Type, and Version
Dynamic device characteristics	Anchor, FreeID, FreeMem, Last, MaxMsgSize, MaxObjSize, Mem, Next, and SharedMem
Misc purposes	EMI, MetInf, and NextNonce

The content-related element types help to specify the content completely. The Format element type is used to define whether the content is character-encoded or binary-encoded. The size of the content can also be specified. Utilize the Mark element type to do special marking, like specifying an email message as unread. Indicating the MIME type of the content is also possible with the Type element. The revision information for the content can be included in the Version element type. Below is an example of an Add command in which all these element types are utilized. The payload data is a calendar event with a size of 145 bytes and using the vCalendar 1.0 MIME type.

```
<Add>
<CmdID>12345</CmdID>
<Meta>
  <Format xmlns='syncml:metinf'>chr</Format>
  <Mark xmlns='syncml:metinf'>final</Mark>
  <Size xmlns='syncml:metinf'>145</Size>
  <Type xmlns='syncml:metinf'>text/x-vcalendar</Type>
  <Version xmlns='syncml:metinf'>20000714T082300Z</Version>
</Meta>
<Item>
  <Source><LocURI>./2</LocURI></Source>
  <Data>
    BEGIN:VCALENDAR
    VERSION:1.0
    BEGIN:VEVENT
    DTSTART:20020602T140000Z
    DTEND:20020602T150000Z
    SUMMARY:SyncML Meeting
    END:VEVENT
    END:VCALENDAR
  </Data>
</Item>
</Add>
```

The SyncML specification allows for substantial flexibility for supporting content-related element types. Only the Format and Type element types are mandatory and must be implemented in SyncML products. The others are optional. Thus, the implementations using Mark, Size, or Version need to pay attention to whether they are supported in the recipient entity.

The element types related to dynamic device characteristics are basically required to enable a Synchronization Protocol-based session between the Client and the Server. The Anchor element type, including Last and Next, is used to transfer sync anchors (i.e. markers)–transferring the anchors allows a conformity check of a previous session, as described in Chapter 5. These anchor-related element types are mandatory for implementations supporting the Synchronization Protocol.

MaxMsgSize can be used to indicate how large a SyncML Message (in bytes) a given device can receive. This information is included in the header of the SyncML Message. The maximum Message size can be specified either once for a session or for every Message transferred. This is one of the very basic element types to be supported for receiving SyncML Messages. Similar to MaxMsgSize, MaxObjSize is used to specify the maximum size of the largest object that can be received in any subsequent response Messages.

The Mem element type with FreeID, FreeMem, and SharedMem includes information about free memory in a device or in a datastore. Support for these element types is optional for the Client, but Client implementations, in general, do support these element types, as they can provide clear benefits. These element types are usually sent to a Server during a SyncML session.

The FreeID element type is used to indicate how many free locations are available for new data objects to be added to a device or to a datastore. For instance, this can be used if a SIM card (Subscriber Identity Module) with a static-sized memory is synchronized with a Server. FreeMem is used when a free amount is indicated in bytes. If the free memory is shared between datastores, then SharedMem should be specified. The utilization of the content in these element types can decrease the amount of data to be transferred between the entities if the receiving entity (e.g. a Client) is close to running out of memory or IDs.

The third category of MetInf element types is used for miscellaneous purposes. In this category, there are three element types: EMI (Experimental Meta Information), MetInf, and NextNonce. The EMI element type can be used to specify nonstandard, experimental extensions. Also, the root element type of the MetInf DTD belongs to this category. Its usage is quite rare, as the child element types can directly be used when utilized within the Representation Protocol. The third element type in this category, `NextNonce`, is needed when challenging for a MD5 digest authentication. Below is an example how to use the `NextNonce` element type within the `Chal` element type of the Representation Protocol.

```
<Chal>
  <Meta>
    <Type xmlns='syncml:metinf'>syncml:auth-md5</Type>
    <NextNonce xmlns='syncml:metinf'>Tm9uY2U=</NextNonce>
  </Meta>
</Chal>
```

From the miscellaneous element types, `MetInf` and `NextNonce` must be supported. `EMI`, as its description already hints, is an optional element type.

WBXML encoding

Like the element types of the SyncML DTD, the `MetInf` element types can be encoded either using clear-text XML or WBXML (binary tokenized) [WBXML01]. When using SyncML over a wireless network such a GSM network, WBXML is commonly used to decrease the amount of data to be transferred.

In both XML and WBXML, the `MetInf` element types can be used within a SyncML Message. Thus, there must be a way to specify the change of a namespace. As seen in multiple examples earlier, the change of the namespace in XML is done according to the XML standard specified by the World Wide Web Consortium (W3C).

In WBXML, specified by the WAP Forum®, there is no standardized way to change the namespace within a WBXML document. Thus, the SyncML Initiative has specified a different WBXML code page for the MetInf DTD. This code page uses the same WBXML code space[3] as the Representation Protocol. The Representation Protocol defines the first code page (0x00) of the SyncML DTD code space to belong to the element types of the SyncML DTD. The second code page (0x01) is then dedicated to the MetInf element types. In this way, these namespaces can be enabled within a WBXML document.

---

3. A code space can include multiple code pages. SyncML has utilized this functionality by using a separate code page for the SyncML DTD and for the MetInf DTD.

Complementary DTD Components

When using the WBXML encoding for the MetInf DTD within a SyncML Message, the code page must be changed in the middle of the WBXML document representing the SyncML Message. Actually, when using an element type from the MetInf DTD, two code page switches are needed. First, the switch for using the MetInf DTD is needed. After the MetInf element type or types are used, then a switch back to the SyncML DTD is required. Below is an XML example showing how to embed the Next element type inside the Data element type of the Representation Protocol.

```
<Data>
  <Next xmlns='syncml:metinf'>20000522T093223Z</Next>
</Data>
```

In a WBXML tokenized form, this would look like:

```
4F 00 01 4A C3 10 "2" "0" "0" "0" "0" "5" "2"
"2" "T" "0" "9" "3" "2" "2" "3" "Z" 01 00 00 01
```

Table 7-2 shows which binary tokens correspond to the text tags. Also, the change of the code page is included.

Table 7-2
Correspondence of Text and Binary Tags

Binary Token	Description
4F	Corresponds to the <Data> tag.
00	Specifies a code page switch.
01	Specifies which code page (now, the MetInf code page) is to be used.
4A	Corresponds to the <Next> tag.
C3	Specifies that an opaque data follow.
10	Specifies the length of the opaque data (0x10 bytes)
"2" "0" "0" "0" "0" "5" "2" "2" "T" "0" "9" "3" "2" "2" "3" "Z"	Specifies the string '20000522T093223Z'.
01	Corresponds to the </Next> tag.
00	Specifies a code page switch

Table 7–2
Correspondence of Text and Binary Tags (Continued)

Binary Token	Description
00	Specifies which code page (now, the SyncML DTD code page) is to be used.
01	Corresponds to the `</Data>` tag.

Device Information DTD

In the SyncML Synchronization Protocol, device- and service-specific information are presented within a document formatted according to the Device Information (DevInf) DTD. This document includes the following characteristics related to a device or a service:

- Hardware characteristics, e.g., the model of a device
- Software characteristics, e.g., the version of the Client software used
- Datastore capabilities, e.g., supported content formats
- Synchronization capabilities, e.g., supported sync types
- Device manufacturer specific information

The DevInf documents are needed for the exchange of the device and service capabilities in the Initialization phase of the Synchronization Protocol. This exchange commonly happens only once for a Client-Server pair. If necessary, the Client or the Server can request to repeat the exchange.

The transfer of the document is similar in nature to any payload data transferred between the Client and the Server. That is, the DevInf document is carried as a MIME object within a Data element type. There are two MIME content types for the Device Information document. The MIME content type of application/vnd.syncml-devinf+xml identifies the XML representation and application/vnd.syncml-devinf+wbxml identifies the WBXML binary representation.

The utilization of the Device Information DTD, in practice, highly depends on the implementations themselves. Clients typically use the device information only to determine a media format and/or datastore names. Clients can even ignore the device information of a Server. In general, Server implementations substantially utilize the device information functionality. Different Server implementations may, however, use the device information quite differently. Some Servers carefully check what is supported in a Client and behave accordingly. For instance, if they realize that a Client does not support some properties of a content type, they can take that into account when synchronizing with that Client. The synchronization services can therefore be dynamically customized for devices of different kinds.

DevInf element types

The element types of the DevInf DTD can be divided into four different categories. The element types can be either related to general characteristics, datastore properties, synchronization capabilities, or miscellaneous purposes. Table 7-3 shows the categories and element types that belong to them.

Table 7-3
DevInf Element Types within Categories

Category	Element Types
Common Information	DevID, DevTyp, FwV, HwV, Man, Mod, OEM, SwV, and UTC
Content Type Properties	CTCap, CTType, DataStore, DataType, DisplayName, DSMem, MaxID, MaxMem, ParamName, PropName, Rx, Rx-Pref, SharedMem, Size, SourceRef, Tx, Tx-Pref, ValEnum, and VerCT
Sync Capabilities	MaxGUIDSize, SupportLargeObjs, SupportNumberOfChanges, SyncCap, SyncType
Miscellaneous	Ext, DevInf, Xnam, Xval, and VerDTD

The element types belonging to the Common Information category mostly specify information that can be beneficially used to associate out-of-band characteristics with a device. For instance, there may be special characteristics related to a specific version of a device model.

The DevID element type specifies the identifier of the source device. For instance, this may be the serial number of the device. The model of a device is specified within Mod. If the OEM (Original Equipment Manufacturer) is different than the actual manufacturer (specified with Man), it can also be specified. The type of the device, e.g. a phone, is indicated in DevTyp. Common Information element types also carry versioning information related to firmware, hardware, and software. Below is an example, showing how a Nokia® 9210 Communicator has used some of these element types.

```
<Man>NOKIA</Man>
<Mod>9210</Mod>
<SwV>256</SwV>
<HwV>301</HwV>
<DevID>IMEI:004400101290920</DevID>
<DevTyp>phone</DevTyp>
```

In version 1.1 of the Device Information DTD specification, the SyncML Initiative added the UTC element type for the Client to specify whether it supports time zones for its applications or not. By including the UTC (Coordinated Universal Time) flag in the device information, the Client can specify that it supports time zones, thus telling the Server to send all application data in relation to the UTC time zone.

The Device Information specification requires the DevID and DevTyp element types to be mandatory. The Server must also be able to receive and process the UTC element type. The other element types are optional. The optional element types vary in their usage, and are not widely used.

The Content Type Properties category and its element types play a very central role in interoperability and the end-user experience, while synchronizing different contents between a Client and a Server. These element types carry the following information:

- Supported content types in the form of MIME types
- Supported receivable and transferable content types for each datastore
- User-displayable name for each datastore
- Maximum storable memory capabilities for each datastore
- Supported properties of each supported content type (i.e. MIME type)

The Content Type Properties element types specify a generic way to represent the supported content characteristic. In other words, they specify which features or fields of a MIME type are supported. The representation of the supported properties can be done in an interoperable way for any MIME type, if the MIME type only needs to be identified. If content type properties and acceptable values and parameter names of the properties need to be specified for the MIME type, they need to be defined in addition to the MIME type itself. The SyncML MetInf specification defines these for MIME types used by calendars, todos, contacts, and memo applications. More about the synchronized content types is discussed in Chapter 11.

There are optional and mandatory element types within the Content Type Properties category. The Client and the Server need to be able to negotiate which content types are supported, which are transferable and receivable, and which properties of the supported content types are actually implemented. The display name and memory information are optional.

The element types of the Sync Capabilities category indicate three types of information. First, the supported Sync Types (from the Synchronization Protocol) are specified inside the SyncCap and SyncTyp element types. For instance, if an entity supports the Sync Types Two-way sync and Slow sync, the support would be indicated in the following way:

```
<SyncCap>
  <SyncType>1</SyncType>
  <SyncType>2</SyncType>
</SyncCap>
```

Second, the MaxGUIDSize element type of the Sync Capabilities category specifies the maximum size (in bytes) of unique identifiers of data items for a given datastore. This element type is only sent by SyncML Clients. In practice, MaxGUIDSize limits the size of the identifiers used by the Server when sending Add operations to a Client. Naturally, the Server may internally use longer identifiers for data items, but when adding items to a Client the Server may have to use temporary identifiers in order to limit the size of identifiers.

The 1.1 version of the SyncML specification introduced features for transferring large objects and for indicating how many changes have been made after a previous synchronization time. These features are optional. Thus, there needs to be a way to indicate whether these features are implemented or not. The SupportLargeObjs and SupportNumberOfChanges element types provide this functionality.

It is clear that the element types of the Sync Capabilities category are crucial to enabling synchronization between a Client and a Server. Therefore, all these element types need to be implemented if they are applicable.

The last category of the DevInf element types, the Miscellaneous category, is dedicated to specifying the following:

- Root element type of the DevInf DTD
- Used version of the DevInf DTD
- Supported nonstandard and experimental extensions

The root and version element types are obviously mandatory for all implementations. The element types used to indicate nonstandard and experimental extensions are only used for proprietary or experimental purposes and the implementations can choose not to support them.

Transport Protocols for SyncML

A SyncML Message is represented as a MIME type. This implies that SyncML Messages can be carried by almost any common transport protocol. It is quite obvious that implementations supporting SyncML cannot support all possible protocols. Thus, they have to decide which protocol interfaces are supported. These decisions might well be based on the following questions:

- Which type of physical connection is desired, wireless or wired?
- Is the physical connection short-range or remote?
- Is there a need to start a synchronization session from a server?
- Is an IP-based or non-IP-based transport connection desired?
- Should the transport connection be standard-based or can it be proprietary?

After answering these questions, Server and Client implementations can decide to support one or more transport protocols. In addition, if a SyncML standard-based transport protocol is desired, the selection can be found among the three transport bindings defined by the SyncML Initiative. SyncML has defined the transport protocol bindings for HTTP, WSP, and OBEX.

In the future, more transports might be utilized by SyncML applications. Some of them may not be standardized if used for proprietary purposes. Future standard protocol alternatives may be related to different messaging protocols and methods, like email or MMS (multimedia messaging).

HTTP Binding

As defined by IETF, the Hyper Text Transfer Protocol (HTTP) [RFC2616] is an application-level protocol for distributed, collaborative, and hypermedia information systems. It is a generic, stateless protocol, which can also be used for many tasks beyond its use for hypertext. A key feature of HTTP is data typing and the negotiation of data representation, allowing systems to be built independent of the data being transferred. Due to this feature, SyncML data (i.e. SyncML Messages) can be conveniently transferred via HTTP.

In the case of SyncML, HTTP (version 1.1) is an ideal protocol for building wide-area distributed systems, such as the synchronization service framework in the Internet environment, whether in the public domain or inside a corporate Intranet. HTTP has proven to be the primary wireline protocol used by SyncML Clients and Servers. HTTP is used over the TCP/IP protocol suite, which is further used over various networks, such as LANs (Local Area Networks). In addition, HTTP can increasingly be used over wireless networks as the bandwidth of wireless networks increases and TCP/IP protocols can be efficiently used over wireless networks.

When using the HTTP binding for a SyncML session, the SyncML Client acts as a HTTP client and the SyncML Server as a HTTP server. The Post method of HTTP is used to carry SyncML Messages from the Client to the Server. The response to the Post method is used to transfer SyncML Messages back to the Client. Because the Client always takes the role of the HTTP client according to the SyncML HTTP Binding specification, the Server alerted sync needs to be done using out-of-band mechanisms like WAP Push. This kind of solution is required, as HTTP does not provide for unsolicited messaging from the server to the client.

Some definite benefits of HTTP are its very wide adoption and the large number of available HTTP products and implementations. In other words, problems related to interoperability and limited support do not exist with HTTP. In addition, there are Open Source HTTP implementations available for both the HTTP clients and servers.

WSP Binding

The SyncML WSP Binding specification defines how to use SyncML over Wireless Session Protocol (WSP) [WSP01]. The WSP binding is based on version 1.1/1.2.1 of WSP. Many mobile terminals are adopting the WSP binding, as they also use WSP for other purposes, such as micro-browsing.

The core of the WAP WSP design is a binary form of HTTP. WSP supports all the methods defined by HTTP (version 1.1). In addition, capability negotiation can be used to agree on a set of extended request methods, so that compatibility to HTTP applications can be retained. WSP provides MIME-typed data transfer for the application layer like HTTP. Compact binary encodings are defined for the standard HTTP headers to reduce protocol overhead.

The main differences between WSP and HTTP are very much related to the fact that the WSP client communicates with the HTTP server through the WAP gateway. From a SyncML Server point of view, it does not really matter whether a SyncML Client uses WSP or HTTP. If WSP is utilized, the WAP gateway makes a translation from WSP to HTTP and the Server sees only the HTTP connection. From a mobile terminal standpoint, the WAP gateway can be described to be an access point to a wired network such as the Internet.

During the translation from WSP to HTTP and vice versa, the WSP client in the device does not interpret all the header information in requests and replies. The WAP gateway partially does that. During the session creation process with a WAP gateway, request and reply headers that remain constant over the life of the session can be exchanged between the client and the server. In addition, the lifecycle of a WSP session is not tied to the underlying transport. A WSP session can be suspended while the session is idle to free up network resources or save batteries. In practice, this means that the physical connection between the WSP client and the WSP server can be disconnected but the logical connection is kept alive.

One of the key features of WSP is WAP Push [WPU01]. This means that a Server can send data in an unsolicited manner to a Client. This is great when the Server wants to trigger the Client to start synchronization. WAP Push can be done over an UDP/IP connection but in general, it happens over non-IP connections like the SMS (Short Message Service). SMS offers an efficient way to send a small amount of data in an unsolicited way from the Server. In the SyncML case, the Server Alert (known as Package #0 in the Synchronization Protocol) is transferred over SMS. The next packages usually go over an IP-connection as the Client establishes the IP connection with the Server after receiving the Server Alert.

OBEX Binding

The SyncML Initiative has selected the Object Exchange (OBEX) protocol [OBEX99] to be one of three specified transport protocols for SyncML. OBEX is commonly exploited over different local media such as IR (InfraRed) and USB (Universal Serial Bus). IrDA® (Infrared Data Association) originally specified it, and due to this, it is also called IrOBEX. The OBEX protocol has been widely adopted by the mobile and PC industry. Most mobile devices (PDAs, mobile phones, etc.) and PCs use it for multiple purposes, such as data-beaming applications.

OBEX is a session layer protocol designed to enable systems of various types to exchange data and commands in a resource-sensitive standardized fashion. The OBEX protocol is optimized for ad-hoc wireless links and can be used to exchange all sorts of objects, like files, pictures, calendar entries, and business cards. OBEX also provides some tools to enable the objects to be recognized and handled intelligently on the receiving side. It is designed to provide push and pull functionality in such a way that an application using OBEX does not need to get involved in managing physical connections. The application only takes an object and sends it to the other side in a "point-and-shoot" manner. This is similar to the role that HTTP serves in the Internet protocol suite, although HTTP is designed more for data retrieval, while OBEX is more evenly balanced for pushing and pulling data.

The SyncML Initiative has defined how to use OBEX over two physical media, IR and Bluetooth™. In both cases, the transfer happens directly over native[4] IrDA and Bluetooth protocols (See Figure 7-4). OBEX could also be run over TCP/IP, but this protocol suite is not commonly found in embedded mobile devices [BIIR01].

In addition to IrDA and Bluetooth, the USB Implementers Forum has defined in an interoperable manner how to use OBEX over USB in the USB Wireless Mobile Communication Devices specification [UWMC01]. Thus, USB can also be used as a physical transport medium for OBEX and SyncML.

Before sending any SyncML Message over OBEX, an OBEX connection needs to be created with the Connect operation. With this operation, the OBEX connection can be dedicated to a SyncML session,

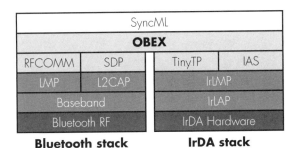

Figure 7-4
Bluetooth and IrDA protocol stacks for OBEX

---

4. In this context, the "native" protocol means a protocol particularly designed for a specific environment, such as the Bluetooth radio environment.

and during this connection phase, necessary parameters for the OBEX connection are also negotiated. After the OBEX connection is made, the Put and Get operations of OBEX are used to send and retrieve SyncML Messages, respectively.

The OBEX role (OBEX client or OBEX server) is independent of the role at the SyncML level. In other words, either the SyncML Client or the SyncML Server can start the OBEX connection and to send SyncML Messages. The entity (SyncML Server or SyncML Client) that starts the OBEX connection is also responsible for disconnecting the connection after the SyncML session is over. For instance, if a PC acting as a SyncML Server has started the SyncML synchronization session with a mobile phone acting as a SyncML Client, the PC needs to disconnect the OBEX connection after sending the final package of a SyncML session to the mobile phone.

8

Security and Authentication

Authentication and security are closely tied together and often taken to mean the same thing. However, they are distinct issues. Authentication is the process of proving who you are via a set of credentials. Security is where you are assured that your communications are private and unaltered.

SyncML Authentication

Even with the most secure transport possible, there is still a need to authenticate the client. After all, it is important to know who is accessing sensitive data. SyncML provides for three authentication layers:

- SyncML® Authentication (different from transport layer authentication)
- Datastore Authentication
- Object Authentication

Each layer's authentication may be overridden at lower layers. For example, the Client authentication may be overridden for a particular datastore, and that may be overridden for a particular object within that datastore.

SyncML Client/Server Authentication

SyncML Client/Server Authentication is the authentication of the Client and the Server. Client/Server Authentication is the most common authentication used in SyncML. This is where the credentials are presented in the `SyncHdr` and are used to authenticate the sender. For simple setups, this level of authentication may be enough. However, in cases where the Client is accessing datastores that contain sensitive information (e.g. payroll datastores), more authentication may be needed.

Datastore Authentication

It is possible that a Client will need access to a datastore that has restrictions on it. For example, a Client may want to synchronize with a corporate datastore that contains the contact information for all of the company employees. It is possible that the user would be granted read-only rights, with only Human Resources people granted read-write rights.

The credentials for this level of authentication would be presented in the `Alert` used to start the synchronization with that particular datastore.

Object Authentication

Object-level authentication is the least used authentication within SyncML. The purpose for this level is to allow individual objects to be accessible to a smaller number of clients than the datastore authentication would allow. For example, an accounting datastore might allow access to the general ledger within the accounting group, and only allow access to the salaries to the Chief Financial Officer and the Human Resources manager.

SyncML Authentication Types

Basic Authentication

Basic Authentication uses only a username or login and a password. These two strings are concatenated with a colon (":") between them (e.g., "billg:snookums"). To keep from transmitting this sensitive data in the clear, SyncML products must encode this string in Base64 encoding. This simple authentication is generally fine for nonsensitive datastores, but may not be adequate for situations where eavesdropping may be a problem, since the encoded string is easily decoded.

An example Basic Authentication for a username of 'userid' and a password of 'password' that has been encoded in Base64 looks like this: 'dXNlcmlkOnBhc3N3b3Jk'. Base64 encoding is described later on in this chapter.

MD5 Authentication

The MD5 algorithm is defined in the IETF RFC 1321 "MD5 Message-Digest Algorithm" [RFC1321]. The MD5 digest is an algorithm that produces a unique 128-bit value when passed in an arbitrary set of data. This data can be text strings or binary values. Note that it is not possible to reproduce the data from the digest–the MD5 algorithm only generates 128 bits of data via a one-way algorithm. The MD5 digest is also useful in determining whether a set of data has been modified in transit, since different data will produce a different digest value.

MD5 Authentication in version 1.0.1

For version 1.0.1 of SyncML, MD5 Authentication is built by creating a temporary string by concatenating the username or login, a colon, the password, another colon, and the recipient-specified Nonce. This temporary string is then run through the MD5 algorithm, producing the MD5 Authentication. The value is placed directly into a WBXML [WBXML01] message, but must be Base64 encoded for XML messages. Readers familiar with HTTP [RFC2616] may note that this is not the same methodology used in the HTTP header.

Here is how the credentials with MD5 will look for a username of 'Bruce2', a password of 'OhBehave' and a Nonce of 'Nonce'. The temporary string will look like this: 'Bruce2:OhBehave:Nonce'.

```
<Cred>
    <Meta><Type xmlns='syncml:metinf'>syncml:auth-md5</Type></Meta>
    <Data>UmkqAAZhiQvXXhfaf/u+OQ==</Data>
</Cred>
```

The algorithm for creating this value looks like the following:

```
MDAuth is resulting authentication value
MD5 is the method of producing a MD5 digest

MDAuth = MD5( username + ':' + password + ':' + nonce )
```

MD5 Authentication in version 1.1

For version 1.1 of SyncML, MD5 Authentication is built differently to allow for separation of authentication from the SyncML processing. The algorithm for creating the MD5 Authentication looks like the following:

```
MD5Auth is the resulting authentication value
MD5 is the method of producing a MD5 digest
B64 is the method of Base64 encoding an arbitrary value
UP is the concatenation of the username and password
MD5(UP) is the MD5 Digest of UP

MD5Auth = MD5( B64(MD5(UP) + ':' + Nonce )
```

The same values as the above example (Bruce2, OhBehave, and Nonce) will produce this XML Cred example (with the data base64 encoded):

```
<Cred>
    <Meta><Type xmlns='syncml:metinf'>syncml:auth-md5</Type></Meta>
    <Data>Zz6EivR3yeaaENcRN6lpAQ==</Data>
</Cred>
```

The MD5Auth value is placed directly into a WBXML message, but must be Base64 encoded for XML messages.

This method is preferred, since it is no longer required to store the password in a datastore accessible to the Sync Engine. The MD5(UP) can be passed directly to an Authentication service for confirmation, and the password is never transmitted over the network.

Base64 Encoding

Base64 Encoding is a simple procedure for turning arbitrary data into standard ASCII (7-bit) data. It also has the nice side benefit of turning simple strings into opaque, or humanly unreadable, data. The process of Base64 encoding takes 24-bit values and treats them as four 6-bit values. Each 6-bit value is used as an index into a printable character table. Table 8–1 contains the Base64 alphabet used for this encoding.

Table 8–1
Base64 Alphabet

Value	Encoding	Value	Encoding	Value	Encoding	Value	Encoding
0	A	1	B	2	C	3	D
4	E	5	F	6	G	7	H
8	I	9	J	10	K	11	L
12	M	13	N	14	O	15	P

Table 8–1
Base64 Alphabet (Continued)

Value	Encoding	Value	Encoding	Value	Encoding	Value	Encoding
16	Q	17	R	18	S	19	T
20	U	21	V	22	W	23	X
24	Y	25	Z	26	a	27	b
28	c	29	d	30	e	31	f
32	g	33	h	34	i	35	j
36	k	37	l	38	m	39	n
40	o	41	p	42	q	43	r
44	s	45	t	46	u	47	v
48	w	49	x	50	y	51	z
52	0	53	1	54	2	55	3
56	4	57	5	58	6	59	7
60	8	61	9	62	+	63	/
Pad	=						

Secure Transport

SyncML has taken the viewpoint that it is preferable to have security handled at the transport layer and authentication at the SyncML Representation layer.

There is no point in creating new secure transports when the best minds in the world are busy working to ensure that the current transports are secure. For example, the secure HTTP (HTTPS) protocol is in use worldwide and is under constant scrutiny to make sure it remains secure.

Of the three transports currently supported within SyncML, HTTP [SHB02] and WSP [SWB02] provide explicit extensions for secure transport (HTTPS [RFC2817] and WTLS [WTLS01]). OBEX [OBEX99] is already a fairly secure transport, as it uses one-to-one media such as RS-232 or IrDA®. Security when using OBEX over Bluetooth [BlIR01] is handled similarly to WSP.

Regardless of the transport used, secure transport guarantees the following:

- A secure "tunnel" between the client and the server exists, disallowing eavesdropping.
- The data is unaltered between the client and the server.

Secure Sockets Layer (SSL)

A brief introduction to SSL is presented here. Greater detail may be found in the various books on security and RFCs.

The SSL Protocol

SSL was originally developed by Netscape® and has been widely adopted for use on the Internet. The Internet Engineering Task Force created a new standard called Transport Layer Security (TLS) [RFC2246] based on SSL. TLS is used in HTTPS and WTLS.

The SSL protocol runs on top of TCP/IP but below higher-level protocols such as HTTP, Lightweight Directory Access Protocol (LDAP) [RFC2451], or Internet Mail Access Protocol (IMAP). SSL uses TCP/IP on behalf of the higher-level protocols. Some side effects of being between the high and low protocols are that SSL: allows an SSL-enabled server to authenticate itself to a SSL-enabled client, allows the client to authenticate itself to the server, and allows both machines to establish an encrypted connection.

Some terms need to be clarified before more is discussed about the SSL protocol:

SSL Server Authentication

Server authentication means a client can confirm a server's identity. SSL-enabled client software can use standard public-key cryptographic methods to check that a server's certificate and public ID are valid and have been issued by a certificate authority (CA) listed in the client's list of trusted CAs. This confirmation is useful in cases where financial information or company secrets are to be stored on a server. In such a case, it is critical that an unauthorized machine cannot act as legitimate server and thereby learn sensitive information.

SSL Client Authentication

Client authentication means a server can confirm a Client's identity. Client authentication uses the same techniques as server authentication. SSL-enabled server software can use the standard public-key cryptographic methods to verify the user's certificates. This sort of confirmation is useful when sending or receiving sensitive data from a user, such as a bank accepting an electronic payment from a user.

SSL encrypted connection

An SSL encrypted connection requires all information sent between a client and a server to be encrypted. This encryption provides a high level of confidentiality–important to any private transaction. In addition to confidentiality, an SSL encrypted connection provides a mechanism to detect tampering–whether the data has been altered in transit.

SSL Cipher Suites

SSL supports a wide variety of standard cipher suites. Note that SSL 2.0 and SSL 3.0 protocols support overlapping sets of ciphers. Here is a list of the more common suites with their owners listed in parenthesis:

- DES–Data Encryption Standard (US Government)
- DSA–Digital Signature Standard (US Government)
- KEA–Key Exchange Algorithm (US Government)
- MD5–Message Digest Algorithm (Rivest)
- RC2 and RC4–Rivest encryption Ciphers (RSA Data Security)
- RSA–public key algorithm (Rivest, Shamir, and Adleman)
- RSA key exchange–key exchange algorithm based on RSA
- SHA-1–Secure Hash Algorithm (US Government)
- Skipjack–classified symmetric-key algorithm implemented in FORTEZZA-compliant hardware (US Government)
- Triple-DES–DES applied three times (US Government)

During initial negotiations, SSL products will select the strongest appropriate cipher for the connection. Decisions about which cipher to use depend on the sensitivity of the data, the speed of the cipher, and the various export laws.

The SSL Handshake

The first part of an SSL session is the exchange of messages, called the SSL Handshake. This handshake allows the server to authenticate itself to the client (using public key techniques), and the server and client to create the symmetric keys used in the normal course of the session. Note that SSL uses both public-key and symmetric-key techniques. The public key is used for the authentication but not for normal session operations, due to the long execution time of public-key methods.

Here is a rough outline of the steps taken during the handshake:

1. Client sends to the server its SSL version, cipher settings, some random data, and any other information the server may need.
2. Server sends to the client its SSL version, cipher settings, some random data, its certificate, and any other information the client may need. If the server needs to authenticate the client, it will request the client's certificate at this time.
3. Client authenticates the server using the information sent in step 2. If the server cannot be authenticated, the session ends.
4. Client creates a Premaster Secret (depends on the cipher being used) using the random data from steps 1 and 2, and encrypts it with the server's public key.
5. Client sends to server the encrypted Premaster Secret and, if requested, the client certificate.
6. If the server receives the client certificate in step 5, it will authenticate the client. If the client cannot be authenticated, the session ends.
7. Both the client and server generate the Master Secret (at the same time), based on the Premaster Secret. The Master Secret is used to generate the session keys–the symmetric keys used to encrypt and decrypt the session information (and encrypt/decrypt the payload data). The session keys can also be used to verify the data integrity.
8. Client sends to the server a message indicating future messages will be encrypted with the session key, and a separate encrypted message indicating the client side is done.
9. Server sends to the client a message indicating future messages will be encrypted with the session key, and a separate encrypted message indicating the server side is done.

More Information on SSL and Certificates

More in-depth information on SSL can be found in some of the IETF's documents, as well as from the major Internet players, such as Sun, Cisco, Netscape, and even the United States Government. For more details on how HTTP works with SSL and TLS, refer to the IETF Web site (*www.ietf.org*). The following RFCs provide detailed information on HTTP, HTTPS, SSL (used in HTTPS), and TLS:

- RFC 2616 (Hypertext Transfer Protocol–HTTP 1.1) [RFC2616]
- RFC 2817 (Upgrading to TSL within HTTP 1.1) [RFC2817]
- RFC 2818 (HTTP over TLS) [RFC2818]

Information on Server Authentication and Certificates may be found in the various X.509 RFCs:

- RFC 2459 (Internet X.509 Public Key Infrastructure Certificate and CRL profile) [RFC2459]
- RFC 2510 (Internet X.509 Public Key Infrastructure Certificate Management Protocols) [RFC2510]
- RFC 2511 (Internet X.509 Certificate Request Message Format) [RFC2511]

9

Device Management

The need for device management (DM) grows ever more crucial as the nature of mobile devices dramatically changes. The complexity of mobile devices is increasing as applications become richer and more powerful. In addition, completely new applications and capabilities are being introduced in mobile devices. These new devices in mass-market environments are used by end-users with minimal technical interest or skills. To hide the complexity of the mobile devices from the end-users and to provide the convenience of seamless functionality of mobile applications, device management technology is necessary.

There are several interest groups in this area in addition to the SyncML® Initiative. Mostly, these groups are associated with standardization organizations preparing standards and specifications in wireless technology environments. Some notable examples are the WAP Forum® and 3GPP™ (3rd Generation Partnership Project). As there are many interest groups and as this technology area is very young in the mobile and wireless environment, even the terminology in the DM area is not yet stable. Other known and commonly used terms for device management are terminal management, user equipment management (UEM), and continuous provisioning.

The SyncML Initiative has a very active role in this area, as it has developed a device management specification (by publishing SyncML DM version 1.1). Two basic reasons justify DM activities within the SyncML Initiative. First, the SyncML Initiative is well suited as an organization to these activities, due to the right mix of member companies. It includes prominent device manufacturers and server software vendors. The knowledge for designing and agreeing on an end-to-end

DM solution and its content exists in this forum. Second, DM has strong similarities to data synchronization. Due to this, major parts of the SyncML data synchronization specification and implementations can be reused for DM purposes.

Technically, DM and data synchronization have a relation that will be discussed further in this chapter. Nevertheless, the main focus of this chapter is on DM itself and SyncML DM technology. This chapter will answer the following questions:

- What is the motivation for device management?
- What can be achieved with device management technology?
- How does the SyncML DM technology work?
- What features are offered by SyncML DM?

Rationale and Overview

In the past ten years, mobile devices such as mobile phones have evolved considerably. The change has been more of a revolution than an evolution. The nature of mobile phones has strongly moved from voice-centric to data-centric. As a result, most new applications and services developed for new mobile phones utilize the data connection capabilities of these devices. In addition to this data-centric movement, mobile applications and services of new types are now more frequently introduced, and end-users are expected to exploit them.

As the nature of mobile devices changes and more new applications are constantly introduced, end-users contend with more feature-rich, complex devices in the mass market. These devices are unavoidably more challenging for users to use and manage, although device manufacturers are putting substantial effort into the design of the usability aspects. Increasing device complexity can lead to end-user frustration with these new features and applications. This in turn can hinder the adoption of these new applications and services.

The possibility that end-users may not adopt and cannot use new mobile applications has caused great concern among various parties, including service providers, operators, device manufacturers, enterprises, and software vendors. All of them see that there is a compelling need to create functionality to help the end-users get familiar with these new applications. DM is a necessary component toward that goal. This need for DM is analogous to what has happened with personal computers. PC end-users are nowadays mostly persons without a technical background. This has commonly led to external parties configuring or maintaining the computers on behalf of the end-users.

Wireless operators especially have a strong interest in DM. This is quite natural, as there are often millions of subscribers in their networks. In this kind of environment, wireless remote DM can substantially decrease the number of customer care calls and increase the usage of data services provided by an operator. Remote DM can be done very efficiently as mobile phones are wirelessly connected to a network infrastructure.

Remote DM over a wireless network (e.g., GPRS [General Packet Radio Service]) is not the only way to manage. DM can also take place over local connectivity mechanisms such as Bluetooth™ or USB™ (Universal Serial Bus). Figure 9–1 depicts different usage environments for DM.

High-level DM functions are very similar to the functions that exist in the traditional desktop system management domain:

- Parameter configuration
- Diagnostics
- Software management

Parameter configuration is intended to manage configurations and settings in devices. For instance, managed settings may enable an application to connect to a service in a network. The diagnostics function enables troubleshooting and performance monitoring of services. Software management is needed for software related operations, such as application installation or upgrade.

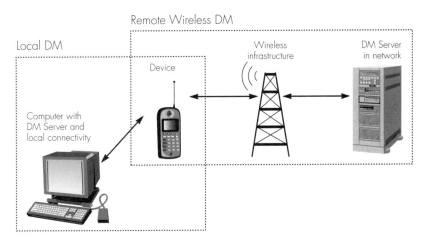

Figure 9–1
DM usage environments

DM technology can be thought of as a functionality that offers services to other applications (e.g. provides them with configuration to be fully operational). Thus, DM cannot be regarded as a standalone application, but as a very important and vital enabler for mobile devices with a large set of mobile applications.

An important characteristic related to the nature of DM is the differentiation of the paying party. Some management operations, such as the update of a service configuration, are typically initiated and desired by the service provider and should not be billed to the end-user. Other operations are triggered by the end-users and should logically be charged to them. For instance, end-users may want to install new applications, for which they will then be billed.

Until the end of 2001, there were no standardized mobile DM technologies that could be used for all the functions listed above. Standards organizations such as the WAP Forum and CDG (CDMA Development Group) had developed specifications for specific usages but they were not generic enough. This has naturally led to the development of proprietary technologies introduced for specific DM functions. Nevertheless, with device management, a standard generic solution is the only acceptable alternative for the long term. Operators, service providers, and enterprises will not be able to efficiently manage very high numbers of devices in their domain if different types of devices and applications require separate management technologies. The SyncML Initiative recognized this deficiency and started to specify a standard, generic DM solution in 2001. As a result, the SyncML Initiative released version 1.1 of the SyncML DM Specification [SDM02] in February 2002.

The SyncML Initiative has specified the management functionality in the mobile domain, but similar work has earlier been carried out in the information technology (IT) domain. For network and desktop system management, organizations such as the IETF (Internet Engineering Task Force) and the DMTF (Desktop Management Task Force) have worked for years on IT systems management. For example, the IETF has released the SNMP (Simple Network Management Protocol) [RFC1157] for managing the diverse computing and communications elements of the modern, distributed environment. Also, the IETF has expanded the capabilities of the SNMP to cover additional MIBs (Management Information Bases). Gradually, this led to the use of the SNMP for the management of far more than just network elements. On the other hand, the DMTF has worked around the technology for building a new generation of managed PC systems and products. The SyncML

Initiative has leveraged the work done by these organizations. It has been inspired by the above specifications, but adjusted them to the domain of wireless and embedded devices.

Benefits to Interest Groups

Device management is a very exciting technology that brings clear benefits to most of the parties associated with mobile devices and mobile services. These parties include the following:

- End-users
- Wireless operators
- Enterprises
- Service providers
- Device manufacturers
- Software vendors

The end-users group consists of the individuals who actually use wireless devices. They may be customers owning their devices or corporate employees using devices provided by enterprises. For end-users, DM itself should be invisible, although it provides direct benefits. For instance, the out-of-box experience can be improved dramatically if a device can be configured automatically on behalf of an end-user. Another good example is remote troubleshooting. With remote troubleshooting ability, an end-user does not need to come back to a retailer or a corporate IT person to fix a possible problem in a device or a service.

Wireless operators and enterprises commonly want to perform the maintenance for their devices. Enterprises do not usually have a large wireless infrastructure but they may own a large number of devices used in various networks. For operators, the situation is quite the opposite. They offer the wireless infrastructure for wireless devices owned by end-users, enterprises, and wireless operators themselves. Despite this difference, wireless operators and enterprises face the same challenge, maintaining all the devices. Naturally, the number of devices is typically lower in the enterprise domain than in the operator domain. Complex devices used in their domains should be used efficiently and smoothly, without problems. Thus, a DM solution to the forthcoming device challenges (complexity, remote diagnostics, error correction, etc.) needs to be available to reduce the costs associated with customer support.

Consumers of the operators and employees of the enterprises commonly use devices of different models and manufacturers. Having a cost-effective way to manage all these devices not only requires DM functionality, but also a DM standard. Having a DM standard such as SyncML DM will make it possible to provide one solution capable of managing the entire pool of devices, regardless of their origin and characteristics.

Service providers, including ISPs (Internet Service Providers), ASPs (Application Service Providers), wireless operators, and enterprises, want their services and applications to be used easily, robustly, and frequently. As a consequence, devices always need to have the right configurations and software to use these services. DM is the technology that will guarantee this.

Device manufacturers mostly gain indirect benefits from DM. DM capability on devices results in good user experiences that users associate with the device brand. In such a situation, end-users will feel that a manufacturer's device is easy to use and that help for using it is always available. In addition to these indirect benefits, a DM standard allows manufacturers to implement only one DM protocol in a device. Hence the overall code size (i.e. footprint) requirement for a device is reduced.

Providing an end-to-end DM solution requires the existence of a counterpart on the service or network side, the DM server software. Software vendors are the companies to provide this type of software. In other words, the demand for DM has created a business potential for these software vendors. The existence of a DM standard is also beneficial to software vendors. A DM standard enables them to focus on value-added features and differentiation rather than being hindered by the need to support multiple protocols.

Usage Models

The SyncML Initiative has identified a set of practical usage models to form the basis for its DM specification. Table 9–1 lists the most common usage models for DM functions. It is worth noting that the list below is not exhaustive. Other usage models can also be supported by the DM functionality.

Table 9–1
DM Usage Models

Usage Model	Example/Description
Provisioning a new device	A brand new device is configured according to the customer's preferences.
Remote service management	After the activation of a service, the configuration for the service is added to a device.
Personal management	A user runs a DM application in a desktop PC. This application enables the management of settings in a device communicating with the PC through local connectivity, such as Bluetooth.
Troubleshooting	A help desk person remotely verifies the operating parameters. If necessary, the help desk person can change the parameters remotely.
Back up and restore	The content of a device is periodically stored on a local PC or a backup server in the network. Later, this content can be restored on the device.
Mass configuration	An operator changes a configuration in all the devices in its networks. For instance, this configuration could be the settings of a GPRS (General Packet Radio Service) access point.
Automatic status reporting	A DM server automatically requests status information from a device, which can be manned (e.g., a mobile phone) or unmanned (e.g., an alarm system in a remote location).
Software download	A new software module is installed, or an installed software module is replaced or deleted on a device.

To see how the usage models can be realized in practice, two common usage models, provisioning a new device and troubleshooting, are introduced with example use cases.

Provisioning a New Device–An end-user buys a new mobile device in a retail store. Store personnel use a device management system to configure the mobile device according to the end-user's preferences. The configuration contains the basic initial activation and connection to a wireless network. In addition, the configuration of premium add-on services, such as browser and email, is performed. The new mobile device and all desired services are fully operational when the end-user leaves the store. Service enrollment takes place immediately at purchase. In addition, similar functionality can be utilized at any time after the purchase.

Troubleshooting—A service provider offers email, WAP, and PIM synchronization services to customers. These services require specific settings in customers' devices. With a DM system, the service provider's helpdesk can check whether all these services are operational in the customer's devices or whether errors have occurred when using them. If errors have happened, the DM system can be utilized to determine and correct the possible problem.

SyncML Device Management Technology

The SyncML Initiative released the first version of the DM specification as a part of the SyncML 1.1 specification. This open technology ensures compatibility between a large number of devices and management servers. The use of SyncML DM has substantial advantages for relatively small and cost-effective devices. It is wireless-friendly and transport-independent. It also provides an extensible, mature, and flexible security model together with independence from a runtime environment.

SyncML DM technology can be divided into the following three logical components:

- Bootstrapping [SBO02]
- SyncML DM Protocol [SDP02]
- Device Description Framework (DDF) [STD02]

Bootstrapping is needed to configure the initial settings on a device. After the bootstrapping phase, only the SyncML DM Protocol is needed for communication between a mobile device and a DM Server. In general, bootstrapping is done over unsolicited message functionality such as WAP Push over SMS (Short Message Service) or OBEX. Bootstrapping includes a separate protocol and additional logic that ensures the necessary level of security for this type of functionality. The Bootstrap Agent (see Figure 9–2) handles this logic.

The SyncML DM Protocol allows management commands to be executed on management objects. It uses the SyncML Message format defined in the SyncML Representation Protocol. In addition, the SyncML DM protocol also reuses parts, such as the authentication procedure, of the functionality of the Synchronization Protocol. Within the DM Protocol, a management object can reflect a set of configuration parameters for a device. Actions upon this type of object might include reading and setting parameter keys and values. Another management object could be the runtime environment for software applications on a device.

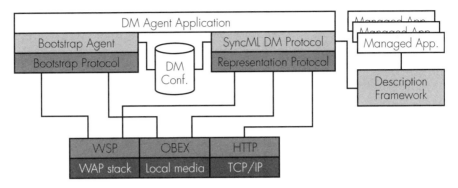

Figure 9–2
SyncML DM architecture

The SyncML Device Description Framework (DDF) specifies the addressing scheme and data structures used by the SyncML DM protocol. The DDF enables device manufacturers and application developers to include all manageable functions in devices in one consistent structure. The DDF actually provides the necessary information for DM Servers to manage the functions in devices of various types. The DDF functionality is quite similar to MIB (Management Information Base) [RFC1213], defined by IETF for the network management.

The manageable applications are located on top of the description framework. A microbrowser, an email client, and a data synchronization application are examples of such manageable applications in a device. They include configuration parameters for the services. In addition, these applications may need to be diagnosed or they may need additional software to provide extra functionality. The characteristics of these applications indicate that external management operations can be beneficial.

Figure 9-2 depicts the whole SyncML DM architecture for a SyncML DM device. In addition to the components described earlier, the transport protocols used in bootstrapping and by the DM Protocol are shown. The DM Agent application provides the necessary application logic and user interface functions. It is built upon the bootstrap components and the DM protocol.

The DM configuration database (DM Conf.), shown between the bootstrap and SyncML components in Figure 9-2, includes all the necessary information for authenticating and communicating with a DM Server. The bootstrap components insert this information and the DM protocol uses the information. The information is needed regardless of the

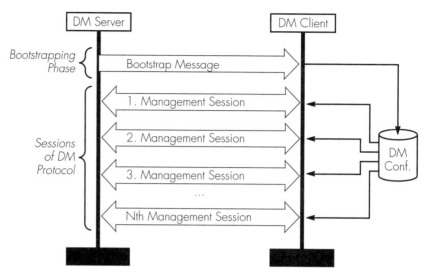

Figure 9-3
Phases of bootstrapping and DM sessions

entity, the DM Server or the DM Client, initiating a management session. The bootstrapping phase and the usage of the DM protocol is illustrated in Figure 9-3. Bootstrapping is needed only once per server. The SyncML DM Protocol is used to run a SyncML DM session whenever needed. As Figure 9-2 implicitly shows, the DDF and the managed applications components are only used after bootstrapping. Those components are used within the SyncML DM Protocol and not during bootstrap.

Comparison with SyncML Synchronization Framework

The similarities of the SyncML DM architecture to the SyncML data synchronization architecture are associated with the usage of the Representation Protocol and with some adopted functions from the Synchronization Protocol. In addition, the DM Protocol uses the same transport bindings as the Synchronization Protocol. Nevertheless, there are some clear differences from the SyncML data synchronization architecture. The main differences are:

- A dedicated management protocol, SyncML DM Protocol, on top of the Representation Protocol
- Device information is represented and used differently
- Stronger security requirements
- Data objects are handled differently

The SyncML DM Protocol is quite different from the SyncML Data Synchronization Protocol, although some functions are reused. A fundamental difference is that the DM Protocol (version 1.1) allows only for the DM Server to send management operations toward the DM Client.

The device information in the SyncML Data Synchronization Protocol is described in an XML document. This is not the case in SyncML DM. The representation of the device info for DM utilizes the SyncML DDF. Thus, all elements of the device info are transferred as individual data items within a SyncML Message. In addition, there are not as many mandated device information elements compared to the number of elements used in SyncML Data Synchronization.

The security requirements within SyncML DM are slightly different and actually stronger. This is motivated by the fact that DM is quite invisible from the end-user point of view and most operations happen in the background. Thus, it is important to eliminate the possibility of denial of service (DoS) attacks or other security threats.

Object handling in SyncML DM is not based on datastores or data items in those datastores. Instead, object handling uses a management tree structure utilizing the DDF that contains all manageable objects.

Bootstrapping in SyncML Device Management

Bootstrapping, also known as initial provisioning, is a prerequisite for starting any "real device management session." Bootstrapping is essentially about provisioning the DM Agent application in a device and possibly configuring some communications settings to enable a transport-level connection with a DM Server. After this operation, a DM session can be started and all other manageable applications can be configured properly. In addition to providing the necessary settings, bootstrapping can logically be regarded as the process that creates the trusted relationship between the DM Server and the DM Client.

Bootstrapping can happen in a number of ways. One method is to use the SyncML DM Bootstrap Protocol to convey the necessary settings to a device. Commonly, this happens over unsolicited transports, such as the WAP Push or the OBEX protocol. The configuration parameters can also be offered within a smart card, such as a SIM (Subscriber Identity Module) card, which is then inserted into the device. A third possibility is to preprogram these settings during the manufacturing process. The last two ways appear to be the most convenient from the end-user perspective, because bootstrapping is already completed by the time a customer buys a device. They have the drawback, however, that if a new DM configuration needs to be added, the smart card or the device itself needs to be reprogrammed.

Many different triggers can spontaneously initialize bootstrapping based on the SyncML DM Bootstrap Protocol. A DM Server can start bootstrapping a device if it detects a new un-bootstrapped device in a network. Alternatively, the DM Server can be ordered to send a bootstrap message to a device. A salesman can do the order at a point of sale where a sales system is connected to a management system. A customer can also do the same thing through a self-service Web site. Thirdly, bootstrapping might also be triggered directly from a terminal. For instance, an end-user sends a specific SMS or calls a specific number to trigger the DM Server. In all these triggering methods, the process and the phases are similar to the ones depicted in Figure 9–4. These triggering methods may utilize many technologies. The SyncML Initiative has not standardized them, as they are implementation-specific and dependent on the underlying technologies, which are not globally unified. The SyncML Bootstrapping specification includes the needed functionality after the DM Server has been triggered.

When sending a SyncML Bootstrap Message, the server has the following two alternatives for formatting the message:

- WAP provisioning bootstrap
- SyncML Message format based bootstrap

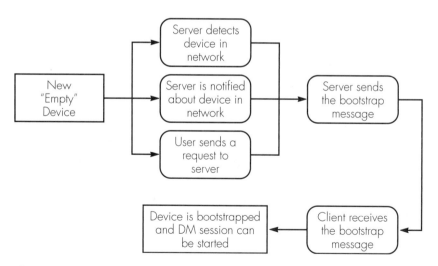

Figure 9–4
Phases of bootstrapping

The alternatives offer similar functionality and are defined by the SyncML Initiative in the SyncML Device Management Specification. As the name of the WAP provisioning bootstrap specification [WPR01] indicates, WAP Forum originally defined this mechanism. The SyncML Initiative simply reuses it. The WAP provisioning bootstrap offers a very complete set of documentation for bootstrapping and can actually be used to provision the WAP browser as such. This bootstrapping mechanism is well designed to work over low-bandwidth transports such as WAP Push over SMS.

The SyncML Message format based bootstrapping utilizes the format defined by the SyncML Representation Protocol. Thus, it reuses existing technology. Implementations that use medium- or high-bandwidth transports might consider using this bootstrapping mechanism. It is worth noting that this mechanism is not very suitable over transports such as WAP Push over SMS, because the message length can be too large.

Bootstrapping is needed to provide the necessary settings to enable communication between the DM Server and the DM Client. The settings include the following information:

- Server addresses, such as a server URL
- Authentication credentials, such as a server ID and a password
- Transport connection settings, such as network configuration for IP connections

An important aspect of bootstrapping is security and how to create a trusted relationship between a server and a device. In practice, this means that the device and the user are able to verify that a bootstrapping message comes from the trusted DM Server. Because of this requirement, the bootstrapping mechanism offers the functionality for authenticating the message itself and also for checking the integrity of the message. Basically, this procedure takes place when a device has received a message. The authentication and integrity check procedure is based on a shared secret, such as a password, which is used to generate an MD5-based digest. When receiving the digest and the bootstrapping message, the recipient can validate the sender and the integrity of the message.

SyncML Device Management Protocol

The SyncML Device Management Protocol is the equivalent of the SyncML Synchronization Protocol. Like the Synchronization Protocol, the DM Protocol can be used over various transport protocols and uses the same SyncML Representation Protocol, although it defines its own usage for the Representation Protocol [SDM02].

As the 'Device Management' term indicates, device management is primarily about managing devices. Thus, the DM Protocol (version 1.1) is designed to provide the DM Server with the functionality of sending commands to the DM Client. The DM Client is not allowed to send any management commands to the DM Server other than `Status` or `Results` commands. Note that this is asymmetric, unlike data synchronization. Within the management commands, the DM Protocol transfers management objects. The management objects are located in the management tree of a device, the structure of which is described by the DDF. The management tree organizes all available management objects in the device as a hierarchical tree structure, where all management objects can be uniquely addressed with a URI.

Managing functionality

The DM Server has two functional sets that are used. Due to this, the management functionality provided by the DM Protocol can be divided into the following two categories:

- Management object functionality
- User interaction functionality

Features belonging to the management object (MO) functionality are dedicated to management objects (MOs) within a device. An example of a MO is a configuration parameter of an email client. Table 9–2 describes the different features under the management object functionality. The table also defines the SyncML command used to realize each feature.

Table 9–2
Features of Object Management Functionality

Feature	Description	SyncML Command
Reading MO content	The server retrieves the content (e.g. parameter value) from the DM Client.	Get
Reading a MO list	The list of MOs residing under a node in a management tree is read.	Get
Adding a MO or MO content	A new dynamic MO is inserted.	Add

Table 9-2
Features of Object Management Functionality (Continued)

Feature	Description	SyncML Command
Updating MO content	Existing content of an MO is replaced with new content.	Replace
Removing MO(s)	One or more MOs under a node are removed from a management tree.	Delete

An example of updating MO content is given below. A DM Server sends the Replace command and the DM Client responds with the Status command. The example illustrates updating the anti-virus definition file in a device.

```
<Replace>
    <CmdID>4</CmdID>
    <Meta>
        <Format xmlns="syncml:metinf">b64</Format>
        <Type xmlns="syncml:metinf">
            application/antivirus-inc.virusdef
        </Type>
    </Meta>
    <Item> <!-- Anti-virus definition -->
        <Target>
        <LocURI>./anti-virus/definition</LocURI>
        </Target>
        <Data>
        <!-- Base64-coded anti-virus file -->
        </Data>
    </Item>
</Replace>
```

The DM Client's response to the command would be the following:

```
<Status>
    <CmdID>4</CmdID>
    <MsgRef>1</MsgRef>
    <CmdRef>4</CmdRef>
    <Cmd>Replace</Cmd>
    <TargetRef>./anti-virus/definition</TargetRef>
    <Data>200</Data> <!-- OK, definition updated -->
</Status>
```

The user interaction functionality of the DM Protocol enables communication with the user of a device. Information about management operations can be provided to the user, or a confirmation for a management operation can be requested from the user. Table 9-3 lists the main user interaction features defined in the SyncML DM Protocol specification. It is worth noting that all the user interaction features are based on the usage of the `Alert` command. These features can be attached to the features of the management object functionality by enveloping those with a `Sequence` or an `Atomic` command. With the `Sequence` command, the order of user interactions and management operations can be defined. The `Atomic` command helps to ensure that all commands are successfully processed.

Table 9-3
Main User Interaction Features

Feature	Description
Display	A notification or additional information about management can be shown to the user.
Confirmation	A confirmation question (yes or no) can be asked of the user.
User input	Input in a text form can be requested from the user.
User choice	One or more selections from a set of options can be requested from the user.

SyncML DM recognizes the need for user interaction for certain management operations and hence provides the capabilities described above. The user interaction features should not be over used, since such overuse is counter to automatic and background device management, the fundamental goal of SyncML DM.

Phases of DM Protocol

A SyncML DM session can be initiated either from the DM Server or from the DM Client. Like the synchronization session, the DM session also has different phases, as depicted in Figure 9-5. The Notification phase is required if the DM Server initiates the session. The Notification phase is usually done over an unsolicited transport protocol, such as WAP Push over SMS. The Setup phase follows the Notification phase. The Setup phase is actually the first phase if the DM Client initiates the session. The last phase, the Management phase, includes an undetermined number of packages, because the DM Server may need to repeat

SyncML Device Management Technology

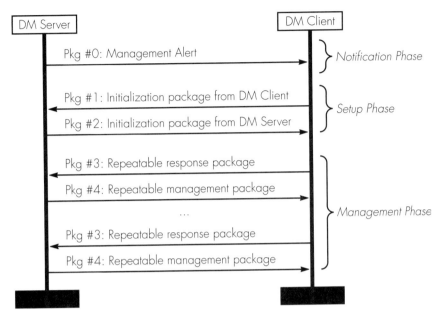

Figure 9-5
Phases of SyncML DM session

Package #4 multiple times. In this case, Package #3 and Package #4 alternate. The Setup and Management phases may use different transport protocols than the Notification phase. They are usually run over transports such as WSP and HTTP.

The Notification phase contains only one Message from the DM Server to the DM Client. It is the management alert for requesting a DM Client to initiate a DM session with a specific DM Server. In addition to the information about the DM Server this Message contains information about the type of management session (e.g. background operation) and about the original entity requesting the management session. The latter can be used to notify the user that an operator has requested the management session.

The Setup phase of a SyncML DM Session includes two SyncML packages, one from the DM Client (Package #1) and one from the DM Server (Package #2). Package #1 is dedicated to the following purposes:

- Sending the device information (manufacturer, model, etc.) to a DM Server
- Identifying the DM Client to the DM Server
- Informing the DM Server about the initiating entity (Client or Server) of the DM session

As a response to Package #1, Package #2 identifies the DM Server to the DM Client. In addition, Package #2 is the first package in which management commands are possibly included. Those commands can belong either to the management object functionality or to the user interaction functionality.

After the Setup phase, the actual Management phase occurs. For the Management phase to occur, Package #2 must contain management commands that require one or more responses from the DM Client. The same condition exists for repeating Package #3 after Package #4. If Package #4 includes management commands that require a response from the DM Client, then Package #3 is repeated. Since the DM Client cannot send any management commands towards the DM Server, Package #3 cannot include any commands.

If the DM Server has received Package #3 and it has no management commands to be sent to the DM Client, Package #4 is used for closing the DM session. In this case, the package is devoid of any management commands.

Security features

There are different security features associated with the DM Protocol, such as authentication, integrity check, and encryption. These features are slightly different for the Notification phase and the Setup and Management phases, as shown in Figure 9–6. The reason is that those phases may be run over different transport protocols.

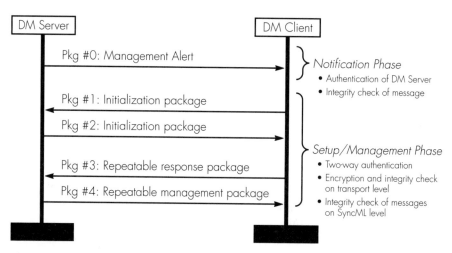

Figure 9–6
Security features associated with SyncML DM phases

As depicted in Figure 9-6, only the DM Server is authenticated during the Notification phase. This authentication guides the DM Client when making a decision whether to start the Setup phase with the DM Server or not. The DM Server cannot authenticate the DM Client prior to the Setup phase.

During the Setup and Management phases, DM commonly uses IP-based transport protocols. The underlying security protocol, such as SSL, TLS, or WTLS, offers the encrypted channel for confidentiality and integrity. Nevertheless, if those security protocols are not supported and the integrity check is needed, the SyncML Initiative has defined a check for this purpose. Integrity of SyncML DM Messages is achieved using an HMAC-MD5. This is a Hashed Message Authentication Code that can be used on every Message transferred between the DM Client and the DM Server if requested to do so by either entity.

Device Description Framework

Mobile devices are very different, both physically and functionally. Nevertheless, it is equally important to be able to manage these diverse devices. Thus, management systems need to be able to handle each individual device even if they have different internal structures and behaviors. To address this issue, the concept of a Device Description Framework (DDF) [STD02] has been introduced within SyncML DM technology. Briefly, this framework provides a way for device vendors to describe their devices so that a management system can understand how to manage them.

The DDF document for a device is created based on the DDF functionality. Having the DDF document of the device enables the DM Server to correctly access the management tree of the device. The DDF document describes how the management tree in the device is constructed and how addressing happens within the device. The DDF document can vary from device to device. As a consequence, the structure of the management trees can also be different. For instance, Device A includes management objects of two types, browser bookmarks and a voice mail number. Device B only includes management object for a voice mail number. In this case, the management trees for the devices could look like they do in Figure 9-7.

The examples in the figure below clearly show that the device management trees for different devices can be different. Although there are some similar functions, they are not necessarily arranged in similar

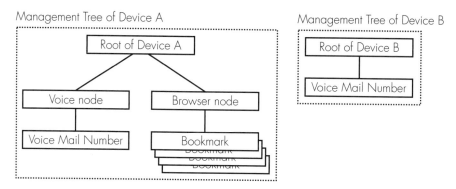

Figure 9-7
Example management trees

ways in different devices. An example of this type of function is the voice mail number.

The DDF document of the device also describes how to address the management objects within the management tree. For instance, if a bookmark is added to the device management tree of Device A, it is added under the Browser node and the management command within a SyncML Message is the following:

```
<Add>
    <CmdID>4</CmdID>
    <Meta>
        <Format xmlns="syncml:metinf">chr</Format>
        <Type xmlns="syncml:metinf">text/plain</Type>
    </Meta>
    <Item> <!-- Bookmark to SyncML web site -->
        <Target>
            <LocURI>./Browser/SyncML</LocURI>
        </Target>
        <Data>http://www.syncml.org</Data>
    </Item>
</Add>
```

Within the management tree, access rights for different DM Servers can vary. In other words, some DM Servers may not have the right to access all the management objects existing in the tree. To enable this functionality, the SyncML DDF includes an access control list (ACL) feature. For example, some Servers may only read the content of a MO whereas other Servers may read and even update the content of the MO.

ACL values for a MO can be controlled and changed by a party that has the specific ACL right for the MO. After having those for a MO, the party's own and others' access rights can be managed. DM Client implementations can choose whether there are one or more parties who are able to modify the ACL values for a MO.

Summary and Next Steps

A key inhibitor in the acceptance and continuous usage of new services, applications, and devices has been the difficulties that end-users experience when trying to correctly configure and maintain their devices. The unfortunate outcome has been poor acceptance and usage of new features, considerable customer support costs for operators and service providers, and poor customer experiences. This dilemma has translated into losses in revenues, increased costs, and lost opportunities.

The DM functionality promises to largely alleviate the difficulties described above. The SyncML Initiative has responded to this demand by creating the SyncML DM specification. The specification covers the protocols and modeling functions, which are needed to have a generic and high-performance DM framework for wireless environments. The framework enables the management of various devices and their continuous seamless operation.

SyncML DM technology is based on SyncML data synchronization technology, but has additional components of its own. Components specific to SyncML DM include the bootstrapping mechanism, the SyncML DM Protocol, and the DDF functionality. SyncML DM also places a stronger emphasis on security, as insecure device management could have devastating consequences.

It is obvious that not all the existing devices will be designed for all DM features. For instance, features of the software management functionality targeted at core software platforms may not be feasible in all products. In general, it is likely that the functions of SyncML DM will be brought to market in the following order: parameter configuration, diagnostics, and software management.

For standardization, the future of DM after the SyncML 1.1 release will likely focus on application management. In practice, this could mean that more standardized object types will be considered for inclusion in the DDF. The second focus area could be APIs that need to be standardized in order to enable the management of third party applications in software platforms. The third possible area of focus could be interoperability tools and testing.

Finally, the reader should keep in mind that DM technology is young (Spring 2002). Unanticipated features and use cases might emerge at any time. Nevertheless, the DM framework and the basic features are available. The products can already take advantage of them and provide important functionality that drives new mobile services and applications forward.

Part III

BUILDING SYNCML APPLICATIONS

10

SyncML API and Reference Implementation

The SyncML® Application Programming Interface (API) and the Reference Implementation have always played an important role in making sure that it is possible to implement the specifications easily and properly, although they are not an official part of the specifications. The SyncML Initiative started the definition of the APIs and the implementation of the reference toolkit in early 2000.

The main reasons that the SyncML Initiative is still investing a lot of time, effort, and money in toolkit development are:

- To prove that the SyncML Specifications are consistent and complete, and that it is possible to implement SyncML, even on small devices with constrained memory and processor capabilities.
- To enable vendors to quickly support SyncML by integrating the toolkit into existing products or by building new products using the toolkit.
- To find flaws and weaknesses in the specifications by enabling vendors to build early SyncML prototype implementations using the toolkit. This allows the Initiative to get valuable feedback from the industry.
- To ensure interoperability of devices by providing a reference.

The first draft version (called "Red") of the toolkit was released at the end of April 2000 to Sponsor-level members of SyncML. Based on this version, they were able to build a set of prototype clients and servers including a demonstration application capable of connecting to Lotus Domino™, as well as to Microsoft Exchange®. This demo was shown to the public at the first SyncML Supporter Summit, in July

2000, held in Los Angeles, California. At the same time these implementations were used for the first validation of the SyncML Representation Protocol [SRP02] and the Synchronization Protocol [SSP02].

A second draft release (called "Yellow") was released in August, followed by a release candidate version (called "Green") in October 2000. The final 1.0 version of the toolkit (called "Gold") was released to the public, together with the 1.0 version of the specifications, during a press event in London, UK on December 6, 2000.

Since then, SyncML has made quarterly maintenance releases of the toolkit available. A version implementing the 1.1 versions of the Representation Specifications for data synchronization and device management was released at the end of March 2002.

Functionality

Figure 10–1 shows the functionality of the toolkit, which was developed in the C programming language in order to allow it to be ported to a wide range of operating systems and platforms. The C language is supported in many mobile devices and network servers. The toolkit provides an easy-to-use API to generate and parse SyncML messages without the need to know the SyncML DTD in depth.

The toolkit does not contain the logic required to handle the Synchronization Protocol and the Device Management Protocol [SDP02].

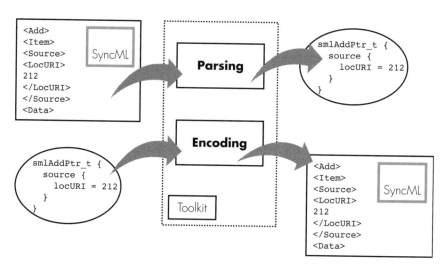

Figure 10–1
Functionality of the Reference Implementation

These are things that an application designer still needs to know and implement. There is a fine line, which the SyncML Initiative does not want to cross, between speeding up the development of SyncML devices by providing a reference implementation and getting in competition with synchronization vendors. Therefore SyncML provides the Reference Implementation Toolkit, which speeds up the development of SyncML Agents, but does not get into the synchronization business itself.

In addition to the reference toolkit, SyncML also provides an open-source sample implementation of an HTTP client and an OBEX client and server. The reference toolkit and the sample transport implementations are also used by the SyncML Conformance Test Suite (SCTS), which is described in Chapter 13.

The following features characterize the toolkit:

- It contains a small and optimized linear parser for the SyncML DTD, and the MetaInf [SMI02] and the DevInf DTDs [SDI02].
- It can encode and decode SyncML messages in XML and WBXML [WBXML01].
- Pluggable components allow for flexible exchange of implementations of specific toolkit components, such as the Encoder/Decoder or the Memory Management.
- Callback functions hand over parsed information to the application.

Currently the toolkit code is developed, tested, and released for the following environments:

- Windows® 98, NT, and 2000
- Linux™ (RedHat Distribution)
- Palm OS® V3.5
- Symbian OS® release 6.0
- Pocket PC

Architecture

The SyncML Reference Implementation consists of the following three layers, as shown in Figure 10–2.

190 Chapter 10 ▸ SyncML API and Reference Implementation

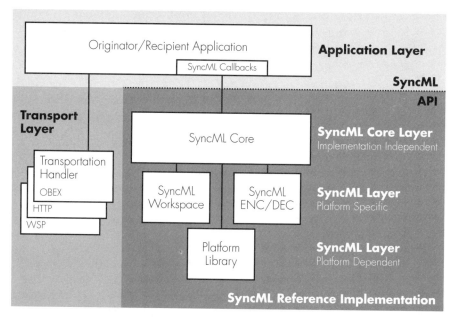

Figure 10–2
Architecture overview

The SyncML core layer implements the SyncML API, which is exposed to the application. The core layer is platform-independent and includes the following three modules, shown in Figure 10–3:

- **SyncML Manager**–This module manages multiple SyncML instances, controls access to the workspace buffer, and switches between sending and receiving messages.
- **SyncML Command Builder**–This module is used to generate SyncML messages. It offers the necessary functions to construct the synchronization packages and command sequences. Appropriate XML or WBXML code is generated.
- **SyncML Command Dispatcher**–This module is used to decode SyncML messages after they are received. It offers functions to interpret SyncML messages and command sequences. XML or WBXML documents are parsed, and callback functions provided by the application are invoked in order to handle each command within the parsed message. The callback functions are responsible for passing the data to the application logic.

The platform-specific layer provides the basic functionality used by the core layer. Modules optimized for particular target platforms can

Architecture

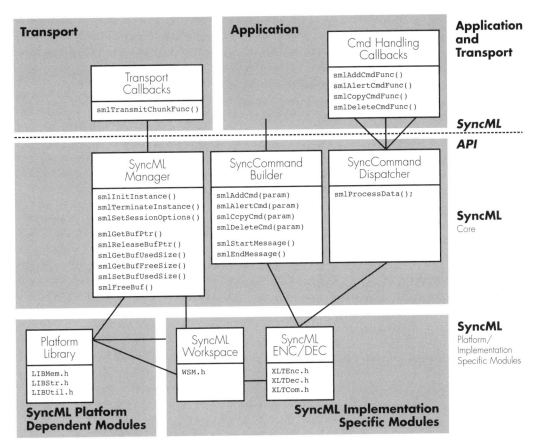

Figure 10–3
Reference Implementation modules

be plugged into this layer to customize the Reference Implementation for specific environments:

- **SyncML Workspace**—The SyncML Workspace handles the memory allocated to incoming and outgoing SyncML Messages. This module is designed as a plug-in, allowing different implementations for different environments. The workspace should be optimized for the particular target platform. The Reference Implementation provides sample implementations for mobile clients, as well as for PCs or servers. A server with plenty of memory available has different possibilities than a PDA or a mobile phone. Nevertheless, the architecture allows the development of a platform independent SyncML workspace if required.

- **SyncML Encoder/Decoder**–The SyncML Encoder/Decoder is designed as a plug-in, encapsulating the parsing and encoding functionality. This approach allows for changing the type of encoding (XML and/or WBXML) and taking platform-specific optimizations into account. Again, it would be possible to develop platform-independent encoders/decoders if required and if an optimization for a particular target is not intended.

The third layer encapsulates the platform-dependent library functions, such as simple memory allocation or string handling. The two layers above use these functions in order to achieve portability.

To use the Reference Implementation and to receive the decoded data, an application has to provide implementations for callback functions. These callbacks handle incoming commands and are called by SyncML while parsing a document. One of these SyncML callback functions implemented by the application, or by the transport module, allows SyncML to trigger the sending of Messages when the workspace runs out of memory.

In order to avoid redundant implementation efforts, the transport layer is not part of the Reference Implementation itself. Instead the application can take advantage of existing transport layer functionality on the individual platforms or use the transport implementation provided by SyncML.

Installation

The Reference Toolkit comes as one zip file for Linux, Palm®, Windows, and Pocket PC and another zip file for EPOC™/Symbian OS. These files contain the complete source and object code for the different platforms. After unpacking the files, a developer should delete the directories that don't apply to his platform. The complete directory tree is shown in Figure 10–4.

Makefiles for the GNU Mingw 32 compiler for Windows and Linux can be found under syncml/src/bld. The Palm OS® version was compiled and tested with the Metrowerks® compiler. The best thing here is to import the code (only the files relevant for the Palm environment, not the Windows or Linux versions of the files) into the Metrowerks IDE.

Initializing the Reference Implementation

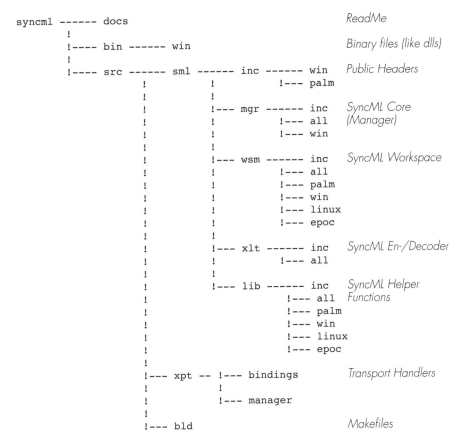

Figure 10-4
Reference Implementation directory tree

Initializing the Reference Implementation

Before using the toolkit to generate or parse any documents, it has to be initialized using the `smlInit` function. Likewise `smlTerminate` needs to be called before the application is terminated to allow the toolkit to free all resources allocated during initialization.

```
SyncMLOptions_t    syncmlOptions;

// Define the maximum amount of memory that
// the workspace should use
syncmlOptions->maxWorkspaceAvailMem = 60000;
syncmlOptions->defaultPrintFunc = NULL;
```

```
//Initialize SyncML
smlInit(syncmlOptions);

// Data processing takes place here
...

// Cleanup all allocated resources
smlTerminate();
```

The above example shows how the toolkit is initialized to a maximum memory buffer of 60,000 bytes usable by the toolkit. If the toolkit should generate tracing or logging information, then the `defaultPrintFunc` needs to point to a callback function allowing storage or display of this information. In this case it is set to NULL to indicate to the toolkit that no tracing information is required.

The examples in the following subsections focus on illustrating how the toolkit should be used. For the sake of readability, these examples make a number of simplifications, which should not be made in production code. For instance, the examples do not check all the return codes from the called functions.

Generating a SyncML Document

To generate a SyncML document, an application first needs to initialize a new instance of the workspace using the `smlInitInstance` command. During initialization, a workspace buffer is assigned. The toolkit uses this workspace buffer to store the generated documents until they are sent or to store incoming documents until they are completely parsed. The application developer could choose to use one workspace for all documents, a different workspace for each document, or one workspace for generating documents and one for parsing documents. The type of encoding that the toolkit should use (XML or WBXML) can be set at the time the instance is initialized with `smlInitInstance` or at a later point in time with `smlSetEncoding`. It is also possible to define a fixed memory size for the workspace buffer. The process is outlined in Figure 10–5.

The following example shows how the instance options are set for generating WBXML documents, assigning the name "MyWorkspace" to the workspace, and setting the size of the workspace to 20,000 bytes. The pointer `&callbacks` points to a structure containing the pointers to the callback functions. The ID of the instance is stored in `instanceID`, which is needed by the application to identify the workspace to be used.

Generating a SyncML Document

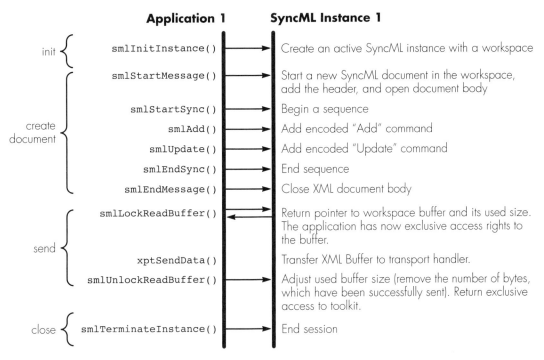

Figure 10–5
Example API sequence for creating and sending a SyncML document

```
// Set instance options and callback functions
instanceOptions.encoding = SML_WBXML;
instanceOptions.workspaceName = "MyWorkspace";
instanceOptions.workspaceSize = 20000;
callbacks.addCmdFunc = &handleAddCmdFunc;
...

// Create instance
rc = smlInitInstance(&callbacks,
                     &instanceOptions,
                     &instanceID);
```

While creating SyncML documents, the application uses straightforward toolkit commands derived from the SyncML DTD [SRP02]: smlStartMessage defines the beginning of a SyncML document. The smlStartSync and smlStartAtomic commands define the beginning of command sequences in the body of a document. For each command specified in the SyncML DTD, there is a dedicated API function, such as smlCopy or smlUpdate. The application is responsible for terminating each message, sync, or atomic sequence with a corresponding smlEndMessage, smlEndSync, or smlEndAtomic call.

The following code snippets show how an Add command is built and added into the XML document.

```
// Allocate needed variables
AddPtr_t       pElem;
MemSize_t      metaLen; sourceLen;
String_t       metaStr; sourceStr;

// Allocate and set structure for Add command
pElem = (AddPtr_t)smlLibMalloc(sizeof(Add_t));
smlLibMemset(pElem, 0, sizeof(Add_t));
pElem->elementType = SML_PE_GENERIC;

// Build PcData element for CmdID "1"
pElem->cmdID = (PcdataPtr_t)smlLibMalloc
   (sizeof(Pcdata_t));
smlLibMemset(pElem->cmdID, 0, sizeof(Pcdata_t));
pElem->cmdID->contentType = type;
pElem->cmdID->length = smlLibStrlen("1");
pElem->cmdID->content = smlLibStrdup("1");

// Build itemList with one item
pElem->itemList = (ItemListPtr_t)smlLibMalloc
   (sizeof(ItemList_t));
smlLibMemset(pElem->itemList, 0, sizeof
   (ItemList_t));

// Build item in itemList
pElem->itemList->pitem = (ItemPtr_t)smlLibMalloc
   (sizeof(Item_t));
smlLibMemset(pElem->itemList->pitem, 0, sizeof
   (Item_t));

// Build target for item. Because it is an Add
// command, the target is not known
pElem->itemList->pitem->target = NULL;

// Add the content
pElem->itemList->pitem->data =
   smlString2Pcdata("Content");

// Add Source element
pElem->itemList->pitem->source = (SourcePtr_t)
```

Generating a SyncML Document

```
    smlLibMalloc(sizeof(Source_t));
smlLibMemset(pElem->itemList->pitem->source, 0,
    sizeof(Source_t));

// Allocate memory for locURI
sourceLen = smlLibStrlen
    ("http://www.syncml.org/servlet/syncit/")+1;
sourceStr = smlLibMalloc((MemSize_t)sourceLen);
smlLibMemset(sourceStr, 0x00, sourceLen);
smlLibStrcpy(sourceStr,
    "http://www.syncml.org/servlet/syncit/");

pElem->itemList->pitem->source->locURI =
    smlString2Pcdata(sourceStr);
pElem->itemList->pitem->source->locName = NULL;

// No flags, because response is required
pElem->flags = 0;
pElem->cred = 0;

// Add the meta element for a vcalendar object
metaLen = smlLibStrlen("<mi:type xmlns=
    'syncml:metinf'>text/vcalendar</mi:type>")+1;
metaStr = smlLibMalloc((MemSize_t)metaLen);
smlLibMemset(metaStr, 0x00, metaLen);
smlLibStrcpy(metaStr, "<mi:type xmlns=
    'syncml:metinf'>text/vcalendar</mi:type>");
pElem->meta = smlString2Pcdata(metaStr);

// Now everything is prepared for the toolkit to
// generate the Add element
smlAddCmd(instanceID, pElem);
```

The sample above uses some helper functions provided by the toolkit, like `smlString2Pcdata`, which generates a Pcdata object from a String. In production code, each of the steps involved in the preparation of the data structures for the Add command (for example, filling the Meta element structure or the target element) should be coded in separate functions. These elements are used by most of the SyncML commands; therefore these separate functions could be reused in most of them without duplicating code. In this example everything was done in one code block to facilitate reading.

After the SyncML document has been created, the application is able to access the workspace buffer containing the assembled SyncML document. This is done using a pointer to the first position of the buffer. The `smlLockReadBuffer` command locks the workspace for exclusive access by the application, and returns that pointer and the actual size of the created document.

The application can now copy the assembled document from the workspace buffer into some outgoing communication buffer. After completing this task, the application must call `smlUnlockReadBuffer` to unlock the workspace and to pass the number of bytes that have been successfully retrieved from the workspace. The toolkit then removes those bytes from the workspace and can reuse the space. After `smlUnlockReadBuffer` is called, the toolkit gets back the responsibility for the workspace buffer. Now this SyncML instance is idle and can be used to perform any other new request.

When an instance is not needed any longer, the instance has to be terminated with `smlTerminateInstance`.

Parsing a SyncML Document

To parse an incoming SyncML document, the application needs to initialize a toolkit instance with an assigned workspace buffer and set the instance options (type of ending, workspace size, and workspace name). For parsing documents, it is especially important that the application registers callback functions for different SyncML commands. The callback functions connect the toolkit to the application. Without the callback functions, the toolkit is not able to pass the parsed data to the application.

```
static Ret_t handleAddCmdFunc (InstanceID_t
    instanceID, VoidPtr_t userData, AddPtr_t param)
{

// Process received data here (e.g. parse user data and
// insert in database )
...
// Free memory allocated to the toolkit
smlFreeProtoElement(param);
return SML_ERR_OK;
}
```

The code snippet above shows what a callback function handling a received Add command could look like. The toolkit is passing the payload data to the application. After the data is processed, the application has to release the memory that was allocated by the toolkit to pass the data.

```
// This sample registers the callback functions

// Define and initialize required variables
InstanceID_t        instanceID = 0;
Callbacks_t         callbacks;
InstanceOptions_t options;

// Set callback functions
callbacks.addCmdFunc   = &handleAddCmdFunc;
callbacks.alertCmdFunc = …
// Continue till all callback functions are set
...

// Set instance options
instanceOptions.encoding      = SML_WBXML;
instanceOptions.workspaceName = "MyWorkspace";
instanceOptions.workspaceSize = 40000;

// Initialize the Instance with options as set
// above
_err = smlInitInstance(&callbacks, &instanceOptions,
         &instanceID);

// Return in case any error occurred
if (_err != SML_ERR_OK) return _err;
```

The callback functions from the previous code snippet are registered with the toolkit at the time the instance is created.

The next step is to lock the buffer for exclusive use by the application, using `smlLockWriteBuffer`. The amount of memory left in the workspace is returned by this function. Another way to do this would be to call `smlGetFreeBuffer`. The application can now store the received SyncML document from the incoming communication buffer into the referenced workspace buffer using `smlLibMemcpy`, a macro provided by the toolkit that hides the platform dependent differences. With `smlUnlockWriteBuffer`, the application unlocks the workspace and returns the control to the toolkit. The size of the data copied into the workspace has to be passed to the toolkit using one of the parameters of `smlUnlockWriteBuffer`.

Now everything is ready to start parsing the data with `smlProcessData`: The toolkit can now parse and dispatch the XML code stored in the workspace buffer. For each identified command, SyncML calls the corresponding callback function supplied by the application.

Chapter 10 • SyncML API and Reference Implementation

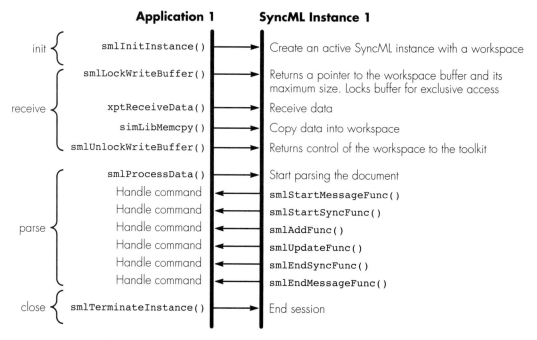

Figure 10-6
Example API sequence for receiving and parsing a SyncML document

This way, the application is able to handle each synchronization command in a customized manner. A parameter of the `smlProcessData` function defines if only one command should be executed at a time or if all commands should be parsed and dispatched at once.

```
// Lock the buffer for exclusive use.
// The pointer to the memory is returned in pdData,
// the available amount of memory in cbDataSize
err = smlLockWriteBuffer (instanceID, &pbData,
         &cbDataSize);

// This is one of the transport functions provided
// with the toolkit.
// Conn is the handle to the connection
// The number of bytes received is returned in
// cbReceived
iXptRc = xptReceiveData (conn, (MemPtr_t) pbData,
           cbDataSize, &cbReceived);
// Unlock the workspace
smlUnlockWriteBuffer (instanceID, (int)cbReceived;
```

This code snippet above shows how the workspace buffer is locked. In the next step, the data received is directly written into the workspace. Finally, the workspace is released and the toolkit informed how many bytes were written into the workspace.

After the smlProcessData function is returned, the application can leave the instance idle or terminate it using smlTerminateInstance.

Communication Toolkit API

The SyncML Reference Implementation comes with a communication API as a complementary component. It helps developers to access the underlying transports using a simple API. It currently supports HTTP [RFC2616] and OBEX [OBEX99]. Originally the support of WSP was also planned, but is not yet implemented. This is due to the fact that WSP would be only attractive to mobile phone platforms, as the supported client platforms can use HTTP instead. Also, the WAP APIs on each platform differ substantially. WSP [WSP01] clients connect to servers using a WAP Gateway, which transforms WSP into HTTP. Therefore servers do not need to contend with WSP, since they get the data via HTTP.

Figure 10-7 shows the architecture of the Communication API. The Communication Manager layer shields the application developer from directly dealing with the different underlying layers. The application just needs to select the appropriate transport and to pass the settings for this transport to the Communication API. This generic approach enables an application to support the transports supported by this API, without writing much extra code for each transport.

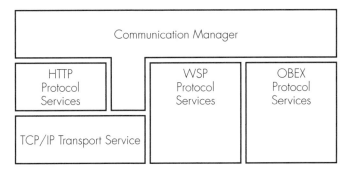

Figure 10-7
Communication API architecture

Using the Communications Toolkit API

When the Communications Manager is started, each available transport is also initialized. Each transport registers itself with a string identifier and a short description. The transport's description could be used by an application to allow the user to select the appropriate communication bearer. A list of available transports is available to the application via the xptGetProtocols call. The description of a particular transport could be retrieved using the xptGetProtocol call, which takes the transport's identifier as an input parameter.

The first step in sending or receiving data is to select and activate a specific transport using the xptSelectProtocol call. With this call, the application specifies whether it is a client or a server for the transport's communication protocol. Transport-specific metainformation is passed by the application, providing the transport with the information it needs to perform its tasks (e.g., host name to connect to, port number to listen on).[1] It is also possible to open several concurrent connections. In server mode, the xptSelectProtocol call causes the transport to start "listening" for incoming requests. Once a protocol is selected, it may be used multiple times to initiate communications. Since some transports open a physical communication path to a particular destination at this time, the application should use a different xptSelectProtocol call for each new destination when in client mode. Each call to xptSelectProtocol returns a handle (called protocol service ID) for the new protocol instance. This handle is used on subsequent calls to identify the protocol instance.

The xptOpenCommunication call is used to initiate communication. In client mode, this typically causes the transport to make a physical connection. In server mode, this causes the transport to wait until it receives an incoming request from a client. A parameter to the xptOpenCommunication call indicates whether the caller will first send a document or first receive one. The simplest approach is to begin by sending a document when the transport is being used in client mode and by receiving a document when the transport is being used in server mode. Each call to xptOpenCommunication returns a handle for the newly created communication instance. This handle is used on subse-

---

1. Ideally, the meta information needed by a transport could be interrogated by an application, so that it might present visual dialogs to the user. This allows the application to be transport-independent, obtaining the meta data from the user or from saved property files. Presently, however, there is no means provided to programmatically describe the transport's meta information, so the application must contain transport-specific settings.

quent calls to identify the instance. Each communication instance may be used to exchange multiple documents.

A document exchange consists of the sender sending a single document and the receiver replying with a single document, as shown in Figure 10–8. Each exchange is surrounded by calls to xptBeginExchange and xptEndExchange. These serve to clearly delineate the boundaries of the exchange and to provide the transport implementation an opportunity to initialize and clean up buffers and other internal management information.

```
xptSelectProtocol(XPT_CLIENT)
    ⎧ xptOpenCommunication(XPT_COMM_SEND)
    ⎪     ⎧ xptBeginExchange()
    ⎪     ⎪ xptSetDocumentInfo()
    ⎨     ⎨ xptSendData() in a loop
    ⎪     ⎪ xptGetDocumentInfo()
    ⎪     ⎪ xptReceiveData() in a loop
    ⎪     ⎩ xptEndExchange()
    ⎩ xptCloseCommunication()
xptDeselectProtocol()
```

Figure 10–8
Sending a document and receiving a document

If the application opened the communication using xptOpenCommunication as a sender, it now should call xptSetDocumentInfo to provide the name and MIME type of the document (i.e., "header information"). With calling xptSendData in a loop, the actual body of the document is provided and sent. It then calls xptGetDocumentInfo to retrieve the name and MIME type of the response document, and then calls xptReceiveData in a loop to read the body of the response document.

In case the application acts as a receiver, it first calls xptGetDocumentInfo to retrieve the name and MIME type of the incoming document, and then calls xptReceiveData in a loop to read the body of the incoming document. After everything is received, it sends a response by calling xptSetDocumentInfo to provide the name and MIME type of the response document, and then calls xptSendData in a loop to provide the body of the response document. Figure 10–9 shows the API sequence for an application first being a receiver and then a sender.

When the communication instance is no longer needed, the application should call xptCloseCommuncation, and when the transport selection instance is no longer needed, the application should call xptDeselectProtocol.

```
xptSelectProtocol(XPT_SERVER)
    xptOpenCommunication(XPT_COMM_RECEIVE)
        xptBeginExchange()
        xptGetDocumentInfo()
        xptReceiveData() in a loop
        xptSetDocumentInfo()
        xptSendData() in a loop
        xptEndExchange()
    xptCloseCommunication()
xptDeselectProtocol()
```

Figure 10-9
Receiving a document and sending a document

The Future

The SyncML Initiative plans to keep the Reference Implementation aligned with the current version of the Representation Protocol. As it is also used in the SyncML Conformance Test Suite, the SCTS might also be a source for future toolkit enhancements.

11
Mobile Devices and SyncML

SyncML® is designed to address the characteristics of wireless networks and mobile devices, such as low bandwidth and a limited amount of memory. Outside the SyncML specifications, there are still many other characteristics relating to wireless devices that need to be considered when designing a SyncML Client. These characteristics are usually linked to the nature of wireless networks, the internal architecture of device platforms, and interoperability.

Implementing SyncML for wireless devices results in a large set of appealing benefits. Mobile and wireless devices are intended for use anytime and anywhere. Thus, data must easily be accessible everywhere at all times. Frequently changed data must be conveniently updated using data synchronization when moving around with a wireless device. SyncML offers a uniform synchronization protocol for wireless devices. This is a great benefit for synchronization solutions, since the majority of wireless device platforms do not allow the installation of third-party software. As a consequence, they cannot support a variety of synchronization protocol interfaces from different manufacturers. Therefore, the synchronization interface frequently needs to be provided by the device manufacturers themselves.

SyncML is an important enabler for mobile applications. As wireless devices become increasingly data-centric, the number of applications requiring data synchronization capability increases. As a result, SyncML will become a mandatory and integral part of mobile device platforms. Before this vision can be realized, the satisfaction and confidence of end-users in mobile data synchronization and

SyncML technology needs to be assured. As a precondition, Client implementations are required to be efficient and interoperable.

Wireless and Mobile Characteristics

Devices and networks operating in wireless environments impose numerous requirements, such as usability and interoperability, on current technologies, applications, and services. These requirements are crucial and should be taken into account when designing and introducing any new technologies, applications, or services to the market. Without addressing these requirements, the new technology may not be accepted by end-users, carriers, service providers, enterprises, and device manufacturers. Each party will have independent reasons for not embracing the technology.

When evaluating mobile technology, end-users commonly look at aspects such as usability, performance, quality, and pricing. In this case, usability means that an application enabling the new technology is intuitive and easy to use without requiring the use of manuals. On the other hand, the performance of the application should not be dramatically weaker than in a wired environment. Typically, users do not tolerate slow response times. Quality is also very important, as mobile applications and terminals are commonly put though physical use that can best be described as "rough." The last, but not least, item in this list is cost. This includes not only the purchase price of the application for the end-user, but also the cost of using it.

For wireless operators and service providers, different aspects are important when introducing a new technology in a mobile environment. Their concerns relate to the usage of overall network bandwidth and the overall cost for enabling a service. It is obvious that the amount of airtime used is important for wireless operators, as more and more operators focus on providing billable services over their network. In that case, the usage of the bandwidth should not consume too much of the limited resources. In addition, operators carefully consider how much they can invest in new mobile technologies without knowing the exact revenue potential.

Device manufacturers also need to address the concerns of end-users. In addition to that, it is crucial to ensure that it is actually feasible to implement a new technology and embed it into a wireless device. This needs to be done in a cost-efficient and competitive manner—which typically means a small footprint with an attractive feature set.

Mobility and wireless technologies offer some definite advantages over wired ones, although there are also limitations. A connection to an infrastructure can be created at any time independent of geographic location. As a consequence, access to data can be established without any major delay or conscious preparation. In practice, mobility gives more freedom to the end-users, and more room to service providers to create new kinds of applications.

Data synchronization, and especially mobile data synchronization, can be used to address many limitations of wireless environments, such as low bandwidth. On the other hand, data synchronization itself can substantially benefit from the advantages of mobility and wide-area wireless networks. Necessary and up-to-date data can always be offered to end-users independent of their location. SyncML, an open synchronization protocol, even expands the benefits of using data synchronization in mobile and wireless environments. Necessary and up-to-date data can be delivered to end-users independent of the mobile devices and network technologies used, as the standard-based approach can be exploited to connect the devices and services.

SyncML Client Architecture and Implementation

SyncML can be designed in multiple ways in different mobile device platforms. Figure 11-1 presents a possible logical architecture, which outlines the key components enabling SyncML Client functionality in a wireless device. It shows components of two types. The common platform components are indicated using *italic* text. The components belonging to the SyncML Client software (Client software) are indicated using **boldface** text. In the figure, several application programming interfaces (APIs) are also depicted to elaborate which types of APIs could be used between the components. This architecture may not be suitable to SyncML Server functionality.

Datastores (DS in the figure), containing the data to be synchronized, are obviously necessary components for the SyncML functionality. For instance, a datastore can contain calendar data, to-do lists, and contact information. In addition to the Client software, there are native applications, such as calendar applications, that present this data to an end-user in the device. The Client software must be able to send and receive SyncML Messages when communicating with a data synchronization Server. For that purpose, the device platform provides one or more transport interfaces (Transport IF in the figure). The transport interfaces offer a

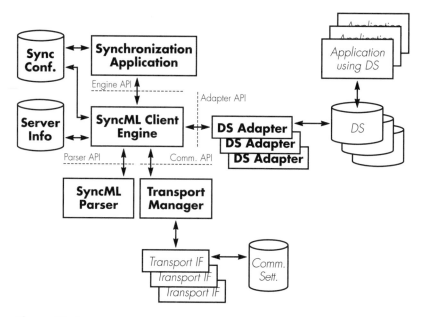

Figure 11-1
Logical architecture of SyncML Client software in a mobile device

data connection over a transport protocol (e.g. HTTP) and a physical transport medium (e.g. GSM/GPRS). The transport components have their own settings for enabling data communication, and these are commonly independent of settings needed for SyncML based synchronization.

As Figure 11-1 depicts, the Client software in a device consists of the following components:

- Synchronization application
- Synchronization configuration datastore
- SyncML Client engine
- Server info datastore
- SyncML parser
- Datastore adapter(s)
- Transport manager

The synchronization application component provides the user interface (UI) and the application logic for starting a synchronization session and for manipulating the synchronization settings. Commonly, this component is dependent on the device platform, the UI style, and the common UI components provided by the device platform. The design of the components can vary appreciably, depending how many different applications, transport protocols, and synchronization Servers the Client software needs to support.

The synchronization configuration datastore in Figure 11-1 contains static configuration information for enabling data synchronization with a Server. These settings are commonly visible to an end-user. The end-user can modify these settings if necessary. Typical settings found in this datastore are the synchronization Server address (e.g. URL), the addresses of the datastores in the Servers, the usernames, and the passwords. There can be other settings depending on the device platform and desired features (e.g. data filtering).

The SyncML Client functionality is implemented into the SyncML Client engine component. In other words, this component executes all the synchronization logic related to the SyncML Client. The logical functions found in the Client engine component are the following:

- Administrating synchronization information related to Servers and datastores
- Managing all synchronized data to and from datastores
- Implementing the needed SyncML handshake procedures for synchronization sessions
- Handling basic conflict resolution when needed

In general, the SyncML Client engine component glues other components together as it interfaces with all the other SyncML related components. The synchronization application component uses the Client engine to execute the synchronization process through the engine API.

The Server info datastore utilized by the SyncML Client engine includes the dynamic information related to synchronization Servers and synchronization events. The information to be stored depends on whether the implementation supports synchronization with multiple Servers or only with one Server. An example of data stored in this datastore is the previous times of synchronization (also called sync anchors) associated with different datastores.

SyncML messages convey information between the SyncML Client and Server. The SyncML Client engine component is responsible for understanding the conveyed information inside these messages. For creating and parsing SyncML messages, the Client engine uses the SyncML parser component. This component can support XML, WBXML, or both. The SyncML Initiative has created an open source parser called the SyncML Reference Toolkit. The toolkit supports both XML-based and WBXML-based Messages. Commonly, the XML or

WBXML parser is used through a Parser API. When implementing the SyncML Reference Toolkit, the SyncML Initiative defined an API based on the C programming language. For details on the SyncML Reference Toolkit, see Chapter 10.

The SyncML Client engine attaches to datastores via the datastore adapter. These adapters enable the usage of different datastores. In other words, the Client engine inserts, modifies, and retrieves data through these adapters. Typically, there is a need for multiple datastore adapters if multiple application types are synchronized. The adapters also hide the differences of datastores from the Client engine. Thus, the Client engine can use one API towards the adapters. In Figure 11-1, this API is called the Adapter API.

The last component related to the SyncML Client software is the transport manager. The purpose of this component is to enable the usage of different transport protocols and physical communication media. For instance, HTTP, WSP, or OBEX can be used as a transport protocol. Beneath them, there are various physical media such as GSM/CSD, GSM/GPRS, TDMA/CSD, Bluetooth™, and WLAN. The supported transport protocols and physical media are dependent on the device platform at which the Client software is targeted.

Main Technical Requirements

The SyncML Initiative designed the specification in such a way that it can be used in mobile and wireless devices. A few important aspects related to the Client software in a wireless device are listed below. This list excludes some items like interoperability, supported applications, and transports.

- Footprint needed
- RAM usage
- Performance
- Usability

The footprint refers to how much space is required for the binary files of a Client implementation. This includes the files added during installation and also the files that may be created by the Client software itself. The acceptable amount of kilobytes that can be reserved for these files is dependent on the specific device platform. For some platforms, only tens of kilobytes are needed, while some other platforms may allow hundreds of kilobytes.

For embedded wireless devices, footprint may be more crucial than RAM usage. This does not have to be the case if the platform enables the execution of multiple applications simultaneously. In any case, the required RAM size of the Client software is very much dependent on the amount of data to be sent or received at a given time. In other words, a large part of the dynamic RAM needed is used to keep a SyncML Message in memory.

Poor performance of Client software applications can lead to two different kinds of results. First, an end-user may be frustrated if a synchronization session takes too long. Second, using the Client software can be too expensive for the end-user if a data connection needs to be maintained for a long synchronization session. The amount of data sent over the air critically affects the performance of the Client software and must be carefully optimized. This is a responsibility of the Client software but also of the SyncML Server with which the synchronization is done. For instance, the slow sync type should not be initiated if not absolutely necessary, and the unsupported properties of application content should not be sent. One aspect of Client software performance is related to internal processes in the device. The internal processes can be related to operations accessing the datastores or the functions establishing the connections.

The usability of the Client software can be primarily divided into three different aspects. The first one is convenient out-of-the-box experience. For an end-user, this means that the first use of the Client software is very easy and trouble-free. This task may not be trivial, as there are settings related to the Client software that must somehow be configured. Different mechanisms (e.g., application wizards, settings sent via a SMS message) can be used for addressing this problem. After the first usability challenge is addressed properly, the continuous usage of the Client software should also be taken into account. Sequential synchronization sessions with a Server must be easy to start. In practice, this means that a synchronization session is initiated without any awkward and time-consuming effort from the end-user. The third area related to the usability of the Client software is the UI functionality during synchronization. The synchronization process and the data connection must be visible to the end-user, as the user may need to cancel or suspend the process in some cases. As mentioned, there can be charges associated with the connection. Thus, these progress- and status-related UI aspects need to be carefully considered.

Minimal SyncML Client Implementation

In a product enabling SyncML data synchronization, the desired functionality in the product generally sets the bounds for the supported features of the SyncML protocols. Nevertheless, when implementing a Client, one may choose to support only the minimal set of SyncML features. In practice, this means that only the mandatory SyncML features are implemented. Those requirements do not define which application types or transport protocols need to be implemented.

Table 11-1 lists the main mandatory features for implementations, or the minimal set of SyncML features. In addition to these, there are many other detailed mandatory features, which can be found in the specifications. It is important to note that the minimal set of SyncML features still offers enough functionality to enable a full Two-way synchronization between a Client and a Server.

Table 11-1
The Main Mandatory SyncML Features for Client

Feature/Function	Special Note
Representation Protocol [SDS02]	
XML/WBXML format	Either one must be supported.
Add command	Only receiving is required, it is not required to send.
Alert command	
Delete command	
Get command	Only support for device information is required.
Map command	Sending is required as a response to an Add command.
Put command	Only support for device information is required.
Replace command	
Results command	Sending is required as a response to a Get command.
Status command	
Sync command	
RespURI element type	Only receiving is required.

Table 11-1
The Main Mandatory SyncML Features for Client (Continued)

Feature/Function	Special Note
Synchronization Protocol [SSP02]	
Two-way Sync Type	
Slow Sync Type	
Change log for data items	Minimally, the support for one Server is required.
Sync anchors	
Identifier mapping	This requires the support of the Map command.
Exchange of device info	Sending is required.
Multiple messages/package	
MD5 digest and Basic authentication	The Client is required to respond to the authentication challenge.

In cases where the Client only supports the mandatory SyncML features, the Server may support some optional features. The Client implementation must tolerate the usage of the optional features, even if it is not able to utilize them.

Mobile Software Platforms

The SyncML Client functionality can be implemented on several different mobile platforms or mobile operating systems. These can also be called mobile software platforms. A mobile software platform with associated device hardware forms a device platform. Some mobile software platforms provide APIs for entities other than the platform vendor itself. For instance, Symbian OS® and Mobile Linux™ are of this type. Some platforms are completely embedded in a vendor's specific device hardware, and only the vendor or its partners can use these platforms to develop applications. Typically, mobile phones and pagers are based on embedded platforms on which it is not possible to install applications after the manufacturing process. Nevertheless, this is also changing, as these platforms are more frequently adopting technologies such as Java™ MIDP (Mobile Information Device Profile). Java-based applications can thus be introduced in these platforms as well.

There is no single answer for how to select a correct mobile software platform for SyncML. The answer depends on many issues, such as what kind of device is desired, what applications are to be supported, what is the region where the application is used, and what are the supported use cases. There are also technical aspects to be taken into account. For instance, an application development company may be specialized in producing software in a specific programming language (PL). Thus, this may be a factor when selecting a mobile software platform for the SyncML application.

Table 11-2 presents a set of major mobile software platforms in alphabetical order. Also, it lists some common characteristics related to them and the devices supporting them.

Table 11-2
Mobile Software Platforms

SW Platform	Primary PL	Multithread support	Native PIM Support	Primary Device Type(s)
Java MIDP	Java	Yes	No	Mobile phone, Smartphone[1]
Microsoft® Pocket PC	C++, C	Yes	Yes	PDA
Mobile Linux	C	Yes	No	PDA
Palm OS®	C	No	Yes	PDA
Symbian OS	C++	Yes	Yes	Smartphone

1. *The term smartphone can include all kinds of communicators, PDA-type phones, and imaging/media phones.*

In addition to these platforms, there are many proprietary embedded mobile software platforms, as discussed earlier in this chapter. The characteristics of these are diverse and are vendor-specific.

SyncML Enabled Applications

SyncML is an important enabler for mobile applications, as a number of different applications need data synchronization capabilities in wireless devices. When creating synchronization services for SyncML enabled wireless devices, it is important to understand which applications are usually supported and how they are supported in these devices.

There are several usage models in which mobile data synchronization is utilized. Four models are addressed below. More synchronization usage models and use cases are introduced in Chapter 3, SyncML Applications.

1. Mobile data synchronization can be used to share information among multiple entities. For instance, calendar data is simultaneously used by an employee and by his/her secretary. Or a group calendar can be stored in a Server with that information shared among many people.
2. The same person as in the first usage model might want data to be available in multiple locations due to usability concerns. When the person is at the office, he wants to access data by using a desktop computer. On the other hand, when he is out of the office, he may only carry a handheld mobile device with him.
3. Backup and restore functionality can easily be provided by a synchronization application. That is, a user may want to back up his data to a network every now and then to ensure that data is not lost.
4. Data synchronization can also be associated with public data distribution. If frequently updated data is distributed into a large number of devices, mobile data synchronization can offer huge benefits in time and bandwidth saved.

The usage models listed above can be applied differently, depending on the application requiring data synchronization. Some of them may not be even suitable or desirable for some applications, while having a high priority for other applications. The usage model is also affected by the target market segment, such as consumer or enterprise.

Application Types

Companies implementing SyncML have started from a limited set of application types and enabled SyncML based synchronization of these applications. For instance, contacts and calendars are supported by most public SyncML Client implementations. Fortunately, the next application types beyond PIM data are already being addressed, as end-users perceive the benefits of SyncML and new applications built on top of SyncML.

Different Client implementations in mobile devices can add support for different application types, but there are a few questions that need to be answered. First, the business case for synchronizing the content of a specific application type needs to be ascertained. Second, it needs to be ensured that there is an interoperable way to synchronize the content type. Without mutually agreed upon data formats (if not standard data formats), it is impossible to achieve interoperability between diverse Client and Servers.

The content format question introduced above has been answered for a set of application types that are commonly used in mobile devices. Those application types are enumerated in Table 11-3. Formats or MIME types associated with these application types are also described. The format describes how the content of an application is transferred in an interoperable way. The timeframe in Table 11-3 represents when it is expected that this application type is or will likely be widely supported by mobile devices. "Now" means that the application type is commonly supported already in 2002. "Near" indicates that the type will be supported within the next 2-3 years. "Far" stands for the distant future or indicates that it is impossible to evaluate when the application support will be enabled.

Table 11-3
Application Types for SyncML in Mobile Devices

Application Type	Format(s)	Timeframe	Special Note
Contacts	vCard 2.1, vCard 3.0	Now	vCard 2.1 is more widely supported by mobile devices.
Calendar	vCalendar 1.0, iCalendar 2.0	Now	vCalendar 1.0 is more widely supported in mobile devices.
To-do list	vCalendar 1.0, iCalendar 2.0	Now	Application type can be integrated into the Calendar application.
Notes	vNote 1.0, Plain Text	Now	Plain Text is more widely supported.
Email	MIME Message (RFC2045, RFC2822)	Near	RFC2045 describes the attachments.
Bookmarks	vBookmark 1.0	Near	
Web pages	HTML, WML, XHTML	Near	WML is used for the WAP 1.x pages.

Table 11-3
Application Types for SyncML in Mobile Devices (Continued)

Application Type	Format(s)	Time-frame	Special Note
SMS	vMessage 1.0	Near	
Generic files	N/A	Near	This can mean any kind of files, such as pictures and documents.
Corporate-specific data	N/A	Far	Proprietary data formats need to be used here, as requirements vary.

The contacts application type is found in most mobile terminals. In mobile phones, it is also called the phonebook application. This application is traditionally used to contain information about personal contacts. In the future, as more contacts are included in mobile devices, the contacts application can be used to handle a wider range of information, like corporate phonebooks or yellow pages. Large datastores of these kinds naturally contain massive amounts of contact information, all of which is changing all the time and must be kept up to date in the device.

Electronic calendars and to-do lists are appearing in more places than just traditional business environments. People utilize them in desktop computers, PDAs, and mobile phones. Data synchronization enables this type of multi-device usage. Electronic calendars have also been introduced in different Web services for both private and public usage. Through these services, all content related to scheduling could also be enabled for mobile phone users. For instance, music concert events or football matches could be synchronized to mobile devices.

In the near future, more application types will be seen in mobile terminals. Some of them are already there, but data synchronization has not yet been utilized for them. Notes, browser, messaging, and file manager are such applications. Using data synchronization with them offers new possibilities to provide additional and different kinds of content to end-users.

As mobile devices evolve, more sophisticated applications will emerge. The emerging applications will likely be more customized for specific purposes, as has happened in PC and Server environments. As this happens, more specific enterprise application types and Relational Databases will be seen in mobile devices. At that time, data synchronization will be essential in order to enable the usage of these applications.

SyncML Requirements of Applications and Datastores

SyncML is content-type agnostic. Nevertheless, there are some requirements for the content to be synchronized using SyncML. In other words, SyncML impacts the application or the datastore of the application residing on a wireless device. By taking this into account in the early phase of design, adding SyncML support for the application is convenient and straightforward.

At a minimum, the application and the datastore implementation of the application desiring the synchronization support are required to provide the following services for the Client software:

- Data items in a datastore of the application are accessible.
- Data items are uniquely identifiable in a datastore.
- Changes of data items and types of changes are detectable.
- Data items can be converted from a native datastore format to a transferable and interoperable content format and vice versa.

The first required service is very obvious. Data items need to be read, modified, and inserted into a datastore. In the Client software architecture presented earlier, this means that the DS adapter operates the datastore as depicted in Figure 11-2.

SyncML technology is designed in such a way that data items are addressed by using local unique identifiers (LUIDs) within a datastore. As a consequence, this is also a requirement for a datastore. All data items in the datastore must be accessible through these LUIDs. Also, when inserting a new data item, a datastore needs to assign a LUID for the item. In practice, the DS adapter uses the LUID as a parameter for operations on the data items in a datastore.

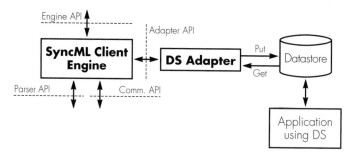

Figure 11-2
The DS adapter accessing a datastore

SyncML Enabled Applications

The third requirement for the datastore of the application is that changed data items must be detectable. This is needed so all data items modified since a previous synchronization session are sent to a Server. There are multiple ways to detect modifications. For instance, the time when a modification has happened can be stored in a datastore. Alternatively, datastores might maintain sequence numbers or dirty bits for the items in the datastore. In addition to the modification detection, the type of modification also needs to be visible. In general, the modification can be an addition, an update, or a deletion. SyncML Clients are only required to indicate whether the modification is an update or a deletion.

It is important to transfer the application content in an interoperable manner. To achieve this, a standard or widely adopted content format is used (e.g. vCard [VCARD21]). This is not necessarily the format in which the applications store the data in their datastores. Conversion to and from the content format is needed when transferring the application content. The conversion can happen in the datastore interface, but it can also happen in the Client software, especially in the DS adapter (see Figure 11-3). If the application uses the same content format for other communication purposes, then this conversion should definitely be offered by the datastore interface.

In wireless devices, there are possibly more requirements for the applications and their datastores to enable the SyncML synchronization. The requirements can be quite application-specific and dependent on the features provided by different application implementations. These types of requirements should be evaluated on an application-by-application or a "mobile software platform"-by-"mobile software platform" basis before the design and implementation of the application is started.

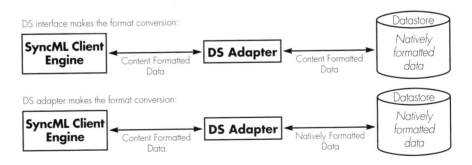

Figure 11-3
Format conversion for data transfer

Summary

Wireless environments introduce some characteristics that are not common in wired ones. These characteristics set the requirements for the Client implementations in mobile devices. In addition, these characteristics also result in real benefits, explaining why SyncML should be implemented in mobile and wireless devices.

The SyncML Client software architectures can vary widely between wireless devices. Such architectures are dependent on the device platforms at which the software is targeted. Nevertheless, there are some common characteristics, which have been discussed in this chapter. In addition, there are common crucial requirements, which need to be considered. Excellent performance and usability are important factors that improve end-user satisfaction.

Having different application types enabled by the Client software can definitely help differentiate mobile devices and their Client implementations. Interoperability should not be forgotten when aiming at such differentiation. Interoperability is not only an issue in designing the SyncML protocol, but must also be taken into account in designing the applications.

The architecture of Client software in a mobile device does not need to be complex. Actually, to fulfill the requirements related to footprint and performance, it must not be complex. A simple architecture implies that SyncML can be implemented easily for mobile devices. This further implies that more and more mobile and wireless devices will support SyncML, and the end-users can substantially benefit from the power of universal data synchronization.

12

The SyncML Server

SyncML® defines the Client and the Server roles for devices participating in synchronization. Chapter 5 elaborates on the differences between these two roles. It is possible for the same device to act as a Client on some occasions and as a Server on other occasions. In practice, however, mobile devices such as PDAs and mobile phones commonly assume the Client role and more resource-rich computers such as PCs and network servers assume the Server role. A device implementing the Client role is afforded certain freedoms in not having to implement all the features of SyncML (see Chapter 11). The asymmetry between the Client and Server roles in SyncML only reflects the inherent, practical asymmetry between a Client and a Server in Client-Server computing. In SyncML, the practical expectations for a Client and a Server are quite different. The Client focuses on ease of use, portability, and memory usage issues. The Server, on the other hand, focuses on an entirely distinct set of issues:

- Managing heterogeneity
- Synchronization analysis
- Performance, scalability, and reliability

A SyncML Server must be able to deal with tremendous heterogeneity, including diverse types of mobile devices and their capabilities. The Server may be required to target and filter data to accommodate the limitations of certain devices. The Server also must contend with various data types, various security mechanisms, and numerous back-end data sources. Clients usually only keep track of changes made to the data on the device, collect and transmit the changes during synchronization, and receive changes from the Server and update the local data accordingly. The Server must perform the bulk of the synchronization analysis—detecting and resolving conflicting updates. In addition, the Server should be able to scale to support thousands of Clients in an efficient and dependable manner.

There are different types of SyncML Servers. The different types may support different groups of users and different classes of applications. Broadly, however, Servers are of the following categories:

- Local synchronization Server
- Service provider Server
- Enterprise Server

Local synchronization Servers typically execute on personal computers offering functions such as PIM synchronization, and backup and restore services for mobile devices. We choose not to focus on such Servers in this chapter, because such local synchronization functionality is not the key driving force for SyncML. The service provider Server typically supports mass-market applications, such as calendars, address books, and email. This type of Server must scale well to many thousands of Clients. The enterprise Server typically supports enterprise applications, such as business email, workflow, inventories, and orders. They usually have richer semantics and higher demands for security and reliability. SyncML Servers are designed in a wide variety of ways. The following describes the architecture of a generic Server. The purpose of this architecture discussion is strictly illustrative and should not be interpreted as a recommended design template.

A Generic SyncML Server

Figure 12-1 illustrates the architecture of a generic SyncML Server. Fundamentally, a Server only needs to implement the SyncML protocols and implement a few applications and data types for compliance and interoperability. It is not necessary for a Server to be structured or be modular in any way. The *form* of the illustrated generic Server simply follows the *function* of any given SyncML Server. The function of a SyncML Server can be partitioned into three main parts.

- Protocol Management
- Sync Management
- Data Management

A Generic SyncML Server 223

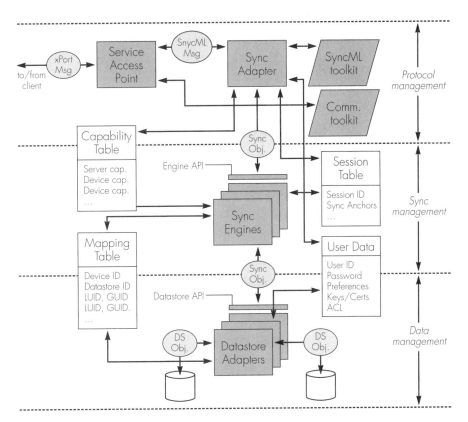

Figure 12–1
The structure of a generic SyncML Server. The components of the Server work together to implement its key functions, Protocol Management, Sync Management, and Data Management.

The Protocol Management part handles functions related to the SyncML protocol itself. These functions include processing and generating messages as per the SyncML Representation Protocol [SRP02], and implementing the handshake sequence as per the SyncML Synchronization Protocol [SSP02]. This part could also handle transport-level authentication and security functions. In addition, in some cases[1] this part may further handle data-level authentication and access control. The Sync Management part primarily handles functions related to maintaining synchronization state and detecting and resolving conflicts. The Data Management part handles functions related to accessing and updating the actual datastores that are being synchronized with

---

1. This is commonly the case for Service Provider Servers.

the Client. In some cases,[2] this part also handles data-level authentication and access control. One advantage of structuring the Server in this way is that this structure insulates the Sync Management logic from changes in the protocols, and the Data Management logic from changes in the Sync Management logic.

Protocol Management

The Protocol Management part implements several functions. It performs transport-level authentication; transforms between the wire transport format and the format specified by the SyncML Representation Protocol[3]; realizes the SyncML Synchronization Protocol; maintains synchronization session information; and transforms between the SyncML format and the different internal formats that may be required by the various Sync Engines. Below we discuss the several components that realize the Protocol Management function.

Service Access Point

The Service Access Point is the primary entry point to the SyncML Server. In many cases, the SyncML Server is a Web server accessed using the HTTP protocol. In such cases the Service Access Point is commonly a Java™ servlet. The access point however, could be a cgi-bin script, or a Visual Basic® script. Messages to the left of the Service Access Point are encoded per their wire transport formats. Messages to the right of the Service Access Point are encoded in the canonical SyncML representation format. The Service Access Point may perform some transport layer authentication and encryption/decryption.[4] In cases where other transports are used (e.g., OBEX [OBEX99]), the Service Access Point could be some custom code accessible via a communication port.

Sync Adapter

The Sync Adapter is the nerve center of the Protocol Management function. First, it may perform SyncML authentication and access control by consulting the user table that stores usernames, passwords, keys,

---

2. This is commonly the case for Enterprise Servers.
3. For the same logical SyncML package, the actual message transmitted varies depending on the transport used or the particular encryption used.
4. The Service Access Point itself may not do this, but its underlying infrastructure could. For example, a servlet typically does not perform authentication and/or encryption per se, but the underlying HTTP Server does.

and certificates. This is especially common in Service Provider Servers, where the Server itself primarily manages the actual datastores. It uses the SyncML Toolkit to transform or generate SyncML Messages to or from "sync objects." The term sync object is used loosely to refer to the internal format that Sync Engines may use. The sync object typically contains synchronization commands and associated data. It is ideal if all Sync Engines associated with the Server can use a common sync object representation. It is possible, however, to have different types of sync objects for different Sync Engines.

The Sync Adapter may also implement the various types of synchronizations in the Synchronization Protocol. It is responsible for conducting the synchronization dialogue with Clients. In addition, the Sync Adapter performs session management for a SyncML session. Using the session table, it keeps track of the progress of a SyncML session and maintains synchronization anchors for Client datastores. It implements failure handling and retries incomplete synchronization sessions.

The Sync Adapter uses the device capabilities table to better control communication with individual Client devices. The capabilities table stores information about individual Client capabilities and also Server capabilities. Client capability information is obtained in the capabilities exchange phase of synchronization. The Sync Adapter, for example, may use Client memory information to limit the size of individual SyncML Messages and decide to send a SyncML Package using multiple Messages.

SyncML Toolkit

The SyncML Toolkit is a set of library routines that assist the Sync Adapter in parsing incoming SyncML messages and generating outgoing SyncML messages. A Reference Implementation of the Toolkit exists, but clearly it is not obligatory for a Server to use the Reference Toolkit. In fact, the Toolkit function could be included in the Sync Adapter itself. Many Servers may find it convenient to use the SyncML Toolkit for the task of parsing and generating SyncML messages. The Service Access Point can use the communication toolkit (part of the SyncML Reference Implementation) to easily process and generate messages specific to a transport protocol. Chapter 10 describes the existing SyncML Toolkit.

Sync Management

The Sync Management function primarily consists of detecting and resolving conflicts arising from concurrent updates to data. Further, Sync Management entails various content adaptation functions, such as filtering and field mapping. For example, filtering enables a Server to send only certain fields of a data record to a particular Client. The key component that performs these functions is the Sync Engine.

Sync Engine

There can be numerous Sync Engines in a SyncML Server. The Sync Adapter invokes the appropriate Sync Engine, depending on the MIME (Multipurpose Internet Mail Extensions) [RFC2045] type of the application or the data involved in synchronization. The Sync Engine component likely implements a Sync Engine API, which the Sync Adapter can use to communicate with multiple Sync Engines in a consistent manner. Note that this Sync Engine API is not part of the SyncML specification. It is simply an internal API that is convenient for a Server to define and use, as it simplifies the implementation of the Sync Adapter. Some Sync Engines may not implement the "common" Sync Engine API. For such Sync Engines, the Sync Adapter will need to communicate using the specific access API provided by the Sync Engine.

The sync object encapsulates the command, the data being synchronized, and certain qualifiers or metadata associated with the data. For example, for synchronizing calendars, the sync object could contain vCalendar [VCAL] data. Along with the data, the object may contain other qualifiers, such as data identifiers on the Client and the back-end datastore, application identifiers, and a datastore identifier on the Server.

The Sync Engine implements all the logic required for sync analysis. Some Sync Engines can be quite generic. For example, there can be one Sync Engine for the personal calendar application that operates on vCalendar data and encodes certain simple rules about conflict detection and resolution. There can be, however, another Sync Engine for the shared family calendar application (see Chapter 3) that operates on the same kind of data, but encodes slightly more complex rules about conflict detection and resolution.

In general, Sync Engines are *configurable*, as shown in Figure 12–2(a). A configurable Sync Engine is driven by policies set by administrators, which are used to resolve conflicts. Policies can be specific to applications or data types. Policies could range from simple ones, such as "Server wins" or "Client wins," to more complex ones, such as "user A's

A Generic SyncML Server

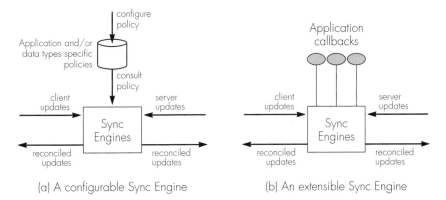

Figure 12–2
Different types of Sync Engines

updates win over user B's" or "Server wins during the day and Client wins during the night," or "conflicting updates are merged." The Sync Engine in this case is essentially a rule engine that is able to execute a set of well-formed rules.

Sync Engines can be *extensible*, as shown in Figure 12–2(b). An extensible Sync Engine allows application programmers to extend the behavior of the Sync Engine by registering callback functions with the Sync Engine. The Sync Engine invokes the callback functions at appropriate times during synchronization for the purposes of conflict resolution. This type of Sync Engine is used in cases where richer application semantics are involved and conflict resolution cannot be encoded in a set of rules. Clearly, a Sync Engine can both be configurable and extensible. For example a provider may choose to implement a Sync Engine for calendars containing some basic rules and policies that can be further specialized for shared family calendars, resource reservation calendars, and business calendars, using application callbacks.

It is common for Sync Engines to perform additional functions to adapt the content for particular Clients. A Sync Engine may use the capability table to determine that a Client only expects a subset of fields present in the back-end data record and to filter the content likewise. Further, a Sync Engine may map between field names used by a Client to and from the field names used by a back-end datastore.

Data Management

The synchronized data ultimately resides in a back-end datastore such as a Relational Database or a file. The datastore itself may actually be an application, such as Lotus Notes® or Microsoft Exchange®, that in turn manages its own physical datastores. A commercial SyncML Server likely contains numerous Datastore Adapters that facilitate the Sync Engine's access to diverse datastores.

Datastore Adapters

A Datastore Adapter converts a generic sync object into a specific datastore object and vice-versa. For example, data for an application such as vCalendar may eventually be stored in a Relational Database. The corresponding Datastore Adapter will convert vCalendar data into the rows of a database table and vice-versa. The application data can also be stored in files. A file Datastore Adapter will convert vCalendar data to and from structures or records and then write or read files. Calendar data is especially interesting, as the datastore can be an application itself. For example, for synchronizing business calendars, the datastore may be Lotus Notes. The Lotus Notes Datastore Adapter will convert vCalendar data to and from the Lotus Notes format and use the Lotus Notes API instead of the JDBC [Ree00] (Java Database Connectivity) API for databases or the file-access APIs offered by the host operating system.

It is desirable from the Sync Engine perspective to communicate with all back-end datastores using a single Datastore API. This ability insulates the Sync Engine from the particulars of the back-end datastore. Note that the Datastore API is not part of the SyncML specification. It is an internal API that Servers may define and use for reasons of convenience. It is reasonable to expect that most Datastore Adapters can support this internal API, simplifying the Sync Engine appreciably. There can, however, be back-end datastores that are not amenable to a "common" API for which the Sync Engine may have to use a specific access API offered by the corresponding Datastore Adapter.

Another key function of Datastore Adapters is assisting in ID mapping.[5] The Sync Engine and the Datastore Adapter cooperate to realize this function. The Datastore Adapter provides back-end identifiers corresponding to insertions made on Client devices that only bear local Client identifiers. The mapping table is used to maintain associa-

---

5. Chapter 5 discusses ID mapping in detail.

tions between local Client identifiers and back-end datastore identifiers. Certain Datastore Adapters, especially in the case of Enterprise Servers, also perform authentication and access control functions.

The Data Management function and Datastore Adapters can be quite complex over and above simple format conversion. In multiple-path synchronization (see Figure 12–5), the Datastore Adapter essentially performs another "synchronization" with the back-end datastore rather than simply reading and writing a back-end datastore.

An Illustrative Dataflow

Let us consider a simple synchronization to understand how actual synchronization may be realized by this generic architecture. Figure 12–3 shows dataflow and the sequence of actions taken by the different components of the SyncML Server while processing an incoming SyncML Package #3 and generating an outgoing SyncML Package #4 from a Two-way synchronization session.

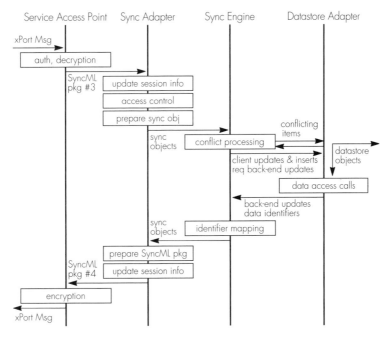

Figure 12–3
The figure illustrates dataflow and the sequence of actions in a SyncML Server processing Package #3 and Package #4 of a Two-way synchronization session.

An encrypted transport message first arrives at the Service Access Point. The Service Access Point authenticates the message, decrypts it, and passes the resulting SyncML Message to the Sync Adapter. This Message happens to be Package #3 of an ongoing synchronization session containing Client updates.[6] The Sync Adapter updates the session table state to reflect the receipt of Package #3. The Sync Adapter then verifies that the user has access rights to the datastores indicated in the Message. In this case, we assume that the Server itself manages access rights to the datastores. The Sync Adapter uses the SyncML Toolkit library routines to convert the SyncML Message into one or more sync objects and invokes a Sync Engine.

The Sync Engine first detects conflicting updates and resolves them. In the process of resolution, the Sync Engine may need to fetch the actual data items using the Datastore Adapter. The Engine also detects new insertions that the Client has made. It further detects that there have been updates on the Server datastore that the Client has not seen. The engine uses the Datastore Adapter to obtain back-end identifiers corresponding to Client insertions and updates the mapping table. The engine also obtains the new updates from the Datastore Adapter. The Engine then returns the results of the conflict resolution and new Server updates as one or more sync objects to the Sync Adapter.

The Sync Adapter uses the returned sync objects to construct a corresponding SyncML Message (Package #4), using the SyncML Toolkit, and returns it to the Service Access Point. The Sync Adapter also updates the session table state to reflect that Package #3 has been processed completely. The Service Access Point uses the SyncML communication toolkit to actually construct the transport message, possibly encrypting the message, and transmit it to the Client.

Data Paths in Synchronization

In designing a SyncML Server, it is important to understand the paths of dataflow from Clients to the actual back-end datastores. In some cases, *all* updates to data pass through the synchronization Server. Thus, from the perspective of the Server, the data has a single path from Clients to datastores, running through the Server itself. This is called *single-path synchronization*. In some other cases, there may be updates to data that do not pass through the synchronization Server. This is called

---

6. If this Package were the first for sync initiation, the Sync Adapter would consult the session table for Sync anchors to detect any failures and initiate slow sync if necessary.

multiple-path synchronization. These two kinds of synchronization occur commonly in real usage scenarios. The synchronization data paths have profound effects on the design of a SyncML Server, especially on the Data Management part of the Server function.

Single-Path Synchronization

Figure 12–4 shows a schematic for single-path synchronization. The individual Clients could be such diverse entities as mobile phones, PDAs, or desktop PCs. The data items being synchronized could be calendar events, email, or music files. Every update, however, must pass through the synchronization Server. This is a common scenario in many Service Provider Servers. Consumers carry mobile devices or desktop PCs and synchronize through the Server. External applications (e.g., an email Server) also update the email datastore through the synchronization Server. The basic underlying philosophy in single-path synchronization is that the synchronization Server is the central focal point of all operations. Hence, all updates must pass through the synchronization Server.

Since all data updates traverse through the synchronization Server in single-path synchronization, the Sync Engine in the Server can store information about observed updates and detect all conflicts readily. In most cases where user intervention is not required, the Sync Engine also

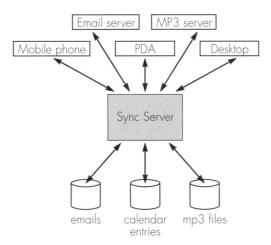

Figure 12–4
Single-path synchronization, where all data updates pass through the synchronization Server.

can resolve the conflicts readily. The Data Management function in single-path synchronization can simply be methods to read and write particular physical datastores, such as files or databases. The generation and management of back-end unique identifiers can also be a function included in the Server (perhaps as part of the Sync Engine), as the Server initiates all back-end operations. In this type of synchronization, usually the datastore has no unique or special role except to offer persistence.

Single-path synchronization lends itself well towards building scalable Servers. All conflicts are readily detectable by the Server. In addition, since all updates pass through the Server, conflicting updates from concurrent synchronization sessions can readily be detected and some updates discarded by the Server without incurring the cost of datastore operations. In most cases, it is also not necessary that the email and mp3 Servers send actual data via the synchronization Server. It is often enough to simply send update information to the Server, such that conflicts can be detected and new Client updates can be generated if required. As indicated earlier, this type of synchronization is commonly observed in Service Provider scenarios. Service Provider Servers may take advantage of all the performance benefits of single-path synchronization to build highly scaleable Servers.

Multiple-Path Synchronization

The philosophy behind multiple-path synchronization is that the synchronization Server is not the focal point of all operations. Instead, the *data itself* is central and the focal point of all operations. Figure 12–5 shows a schematic for multiple-path synchronization. This kind of synchronization is common when existing applications are extended for mobile or disconnected users. For example, it is commonplace to update business inventory data from a traditional desktop PC. In the mobile age, users carrying mobile devices would like to update the same data. The updates generated by the mobile users pass through the synchronization Server. The legacy interface to inventory data still remains, as there is no compelling reason to force the legacy applications to pass through the synchronization Server to fit into the single-path model. Business email also falls in the same category. People are already accessing business email (e.g. Lotus Notes) from connected desktops and directly using email from their desktop applications. The mobile users pass through the synchronization Server but the traditional users have no good reason to do so. It is also not advisable to compel the traditional users to pass through the synchronization Server, as that would entail

Data Paths in Synchronization   233

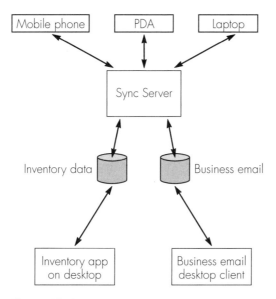

Figure 12-5
Multiple-path synchronization, where not all data updates pass through the synchronization Server.

sacrificing the traditional user experience in favor of the mobile user experience. Changing large numbers of useful and popular traditional applications also may not be economically feasible. These kinds of scenarios are common in the enterprise domain, as evidenced by the examples offered. To gain acceptance in an enterprise, it is important that the SyncML Server support multiple-path synchronization.

Data Management Issues

The primary difference in single-path and multiple-path synchronization is in the complexity of the Data Management function. The Data Management function in single-path synchronization is straightforward. Figure 12-6(a) shows the general structure of such Data Management. Since all updates are guaranteed to pass through the Sync Engine (inside the synchronization Server), the Sync Engine itself can maintain an update history. The update history shown in the figure can also be stored in persistent storage, such as a Relational Database (like the data itself). When the Sync Engine receives Client updates, it can consult the update history to detect conflicts. In the simple case, it can resolve all conflicts (possibly by invoking application code on the Server) and send

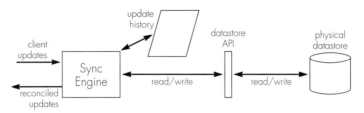

(a) Data Management for single-path synchronization

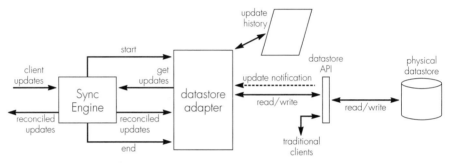

(b) Data Management for multiple-path synchronization

Figure 12-6
The different kinds of Data Management required to support single-path and multiple-path synchronization.

reconciled updates to the Client, as well as make necessary changes to the physical datastore using the datastore API. In some cases, the Sync Engine may only mark the conflicts and ask the Client (user) to resolve them. The results of the resolution can be simply written back to the physical datastore in a later phase of a synchronization session. The datastore API is usually simple. The API normally only translates the canonical internal object representation of data items to the representation used by the physical datastore.

Data Management for multiple-path synchronization is more complex. A general structure for such Data Management is shown in Figure 12–6(b). The complexity arises from the fact that the Sync Engine is not aware of all the updates made to the physical datastore. Normally, the Server needs to implement Datastore Adapters specific to various different kinds of back-end datastores, such as Relational Databases, Lotus Notes, or Microsoft Exchange. The Datastore Adapters enable the Sync Engine to simply request updates that have been made to the datastore since some previous time. The Sync Engine then detects and resolves conflicts with the Client updates (in the simple case) and sends reconciled updates back to the Client, as well as to the Datastore Adapter. The Datastore Adapter sends the reconciled updates to the back-end store using the datastore API.

The Datastore Adapters must be able to track changes to the back-end datastores and maintain an update history (possibly in persistent storage), even though certain updates to the physical datastore may flow not through the Datastore Adapter, but directly through the datastore API. Different physical datastores support different mechanisms by which a program (such as the adapter) can track updates made to the datastore. For example, Relational Databases support triggers or database log-analysis software to track updates. Lotus Notes Servers also support similar functions. Clearly, the Datastore Adapter is quite specific to the back-end datastore, and a synchronization Server may have multiple such adapters for various supported back-end stores.

Changes to back-end stores may be made concurrently when a synchronization session is in progress. The Datastore Adapter must be able to reconcile the concurrent updates with the updates pertaining to a synchronization session. The Sync Engine typically needs to mark the start and end of a synchronization session with the Datastore Adapter such that the adapter can track updates made to the back-end store concurrently during synchronization. Some back-ends may allow locking the physical datastore from concurrent updates while a synchronization session is in progress. Such restrictions, however, are not applicable in general. The Datastore Adapter must often support mechanisms to queue concurrent back-end updates during synchronization and reconcile them later, perhaps with the next round of synchronization.

Implementing sophisticated Datastore Adapters for multiple-path synchronization is complex. The adapters actually could be the bulk of the work involved in implementing the synchronization Server. For high performance, the adapters often cache updates, and the cache must be kept synchronized with the back-end datastore. Certain adapters are even more complicated due to semantics of data. For example, certain write operations to Relational Databases may fail because they violate the integrity constraints of the data (see constraint violation in Chapter 1). The adapter and the associated Sync Engine must have the means to handle such errors even after the Sync Engine has processed conflicts.

Functional Expectations from a SyncML Server

A production SyncML Server is expected to have certain functional abilities that are commonly more than the minimal requirements for compliance and/or interoperability. The expectations include support for the following:

- Many device types
- Numerous data types
- A few different Sync Types
- Various authentication schemes
- Diverse types of back-end data sources
- Multiple transport protocols

Many types of devices may wish to synchronize with a SyncML Server. Devices vary in capabilities such as communication bandwidth and memory size. A SyncML Server needs to be aware of device capabilities and adjust its behavior accordingly. For example, depending on Client memory conditions, the Server may decide to partition SyncML Packages into multiple Messages. A Server also needs to share its capabilities with certain devices (e.g., supported applications and data types) for devices to make better decisions about synchronization. Servers use the Initialization phase in the Synchronization Protocol to discover device capabilities and store these capabilities in an internal table.

With diversity of devices comes diversity of applications and diversity of data that the applications use. For compliance and interoperability, the SyncML Server already needs to support multiple common PIM data types. Other desirable data types include relational data, XML and HTML data, and binary data, such as music and pictures. An important issue in supporting multiple data types is the ability to support conversion of the data types to and from their external "standard" representations to "internal" or "back-end" representations. For example, there are many commercial calendar systems that do not export APIs that use the vCalendar format. If the Server wishes to allow mobile Clients to synchronize with such systems, it has to provide appropriate converters. As mentioned earlier, Servers may be required to perform filtering and field mapping for various data types and Clients. Sometimes, the display capabilities of mobile devices are limited. Certain forms of data (e.g., pictures) may need to be *transcoded* (e.g., from color to black-and-white) before they can be sent to the device. To enable end-to-end solutions acceptable to users, Servers must implement such transcoding functions or leverage existing transcoding libraries.

SyncML Servers may need to support all the Sync Types in the specification. The types include One-way synchronization with only Client updates communicated to the Server, One-way synchronization with only Server updates communicated to the Client, and Two-way synchronization. Moreover, the Servers need to support all of the types where the Server is initiating the synchronization using a Server Alert message to a Client that is connected (by a mobile phone, for example). They also need to handle cases where some devices may choose not to respond to Server alerts. The Servers also need to support "slow synchronization" upon detecting certain failures, where the contents of entire datastores are exchanged instead of only updates.

Applications require varying levels of authentication. Some applications only require basic username-password authentication, while other applications require more elaborate authentication using message digests or Certificates. In some scenarios, authenticating the user with the SyncML Server is enough. In other scenarios, the user needs to be authenticated to individual datastores. In yet other scenarios, the user may need to be authenticated for individual object access in a given datastore. The SyncML Server should be able to support most (if not all) of these authentication levels.

A SyncML Server, especially an Enterprise Server, should be able to support multiple back-end datastores. This requirement not only includes supporting back-end specific formats but supporting access to back-ends using the specific APIs offered by the back-end datastores. If a Server stores some data in a Relational Database, it would need to use the standard JDBC or Open Database Connectivity (ODBC) [San98] APIs. If the Server chooses to store data in the file system, it needs to use the file-access APIs of the underlying platform. In some cases, the back-end datastore may actually be an application that exposes its own API, such as Lotus Notes[7] or Microsoft Exchange. In such cases, the SyncML Server needs to use the APIs provided by these applications. As observed in the previous section, in single-path synchronization, it is a relatively straightforward matter of invoking the correct back-end APIs for data access. In multiple-path synchronization, supporting diverse back-end stores is more complex, as specific Datastore Adapters may need to be implemented that need more complex function than only data access.

---

7. Lotus Notes is the name for the end-to-end application. The Notes Server is actually called Lotus Domino. At many points in this chapter, we implicitly refer to the Lotus Notes Server and hence actually to Lotus Domino.

SyncML Servers also need to support multiple transport protocols. It will be common for SyncML Servers to be accessed by HTTP Clients or by WAP™ (Wireless Application Protocol) Clients coming through a WAP gateway. Although one can argue that the WAP gateway hides the fact that the Client is a WAP Client and converts the WAP messages into HTTP messages, the Server can still benefit by knowing the eventual communication protocol that the Client uses. For example, WAP gateways have message size limitations that may compel the SyncML Server to partition Packages into multiple Messages that it otherwise would not have to for an HTTP Client.

Performance, Scalability, and Reliability

A SyncML Server will likely serve thousands or tens of thousands of Clients. SyncML compliance and the above functional characteristics are necessary but not sufficient attributes of a production SyncML Server. A Server should offer high performance, be scalable, and be reliable. This section discusses a number of means to design a SyncML Server to help achieve the above goals.

Exploiting SyncML Characteristics

In Chapter 4, we discussed certain characteristics of SyncML data synchronization that enable the building of scalable Servers. They are the following:

- Batch operations
- No constraints on ordering of operations
- No transactional guarantees

SyncML allows operations on datastores to be batched in one SyncML Package. A SyncML Server must take advantage of this batching, as many back-end data sources have appreciable connection overheads that can be effectively amortized by batching operations on the datastores. In addition, the SyncML Server can attempt to batch operations across many concurrent synchronization sessions. Batching operations across concurrent sessions enhances overall performance. Further, it enables the Server to detect and resolve conflicts caused by concurrent updates before incurring the cost of datastore operations. Most commands in SyncML place no constraints on the ordering of operations. This facilitates building highly concurrent, multithreaded systems where numerous threads process parts of the overall task and do not have to co-ordinate to preserve any order. Most operations in SyncML also do not offer transactional guarantees. This implies that the Server threads do not have to wait until back-ends actually confirm that operations have succeeded.

Exploiting Back-End Characteristics

Batching back-end operations, as discussed above, is the first step toward exploiting back-end characteristics. For administrative reasons, it is often found that the back-end database is resident on a machine other than the synchronization Server. In such cases, back-end operations incur the additional cost of network transfer. Several back-ends, however, have in-built replication support (e.g., Lotus Notes). The synchronization Server should take advantage of back-end replication support whenever possible by using a local replica of the back-end datastore and replicating with the actual back-end datastore at certain intervals.

In multiple-path synchronization, it is especially critical to be able to exploit back-end characteristics when building the Datastore Adapters. It is sometimes difficult to keep track of changes made to back-end data (by non-sync applications). Several back-end stores allow means for applications to be notified of changes. For example, Relational Databases allow triggers, which can notify the Datastore Adapter when changes are made to the database. Certain Relational Databases allow applications to track changes by allowing APIs to access the database operation log. Depending on the particular database used, one mechanism may be more efficient than the other. Non-Relational Databases, such as Lotus Notes, often allow applications to register agents that can notify applications of changes. The Server implementation must carefully choose between alternative mechanisms available for different back-ends on a case-by-case basis.

Exploiting Application Characteristics

Large classes of applications that are commonly considered data synchronization applications have special characteristics. Some applications are *read-only*, where Client devices only read Server data and never change it. An insurance agent downloading daily rate quotes is a read-only application. Some applications are *read-write* applications, which do not generate any conflicts by design. For example, if one uses a personal email application from a single mobile device only (a common case), emails are read and written but a conflict never arises. When synchronizing such applications, the Sync Adapter can make direct calls to the datastore APIs without incurring the cost of invoking the Sync Engines.

Application *usage pattern* is also an important factor to consider. Many applications will generate a high load only during certain times of the day. If the Server is able to anticipate such high load conditions for an application, it can establish anticipatory connections to back-ends, and it can prefetch and cache back-end updates. These steps will likely reduce the overall latency observed by the application. Servers will need to maintain application usage statistics and reconfigure parameters based on such statistics.

Effective Use of Concurrency and Asynchrony

Experiences in system design show that a production SyncML Server must be able to exploit concurrency effectively. One can imagine designing the Server as one complex, monolithic procedure that implements all the functions of Protocol Management, Sync Engine, and Data Management. Such a system is not conducive to low latency and high performance, as an arriving Client request will need to wait until every earlier request is completely processed. In addition, such a monolithic system is difficult to extend for new data types and applications.

Figure 12–1 offers a tacit suggestion that a SyncML Server can be built in *pipelined stages*. Each stage of the pipeline offers concise functions such as authentication, SyncML Message parsing and/or generation, conflict detection/resolution, and data access operations. Pipelined stages allow a preceding stage of the pipeline to service the next Client request while the current stage handles the current request, increasing the throughput of the system dramatically. Multiple concurrent threads can realize each pipelined stage. Ideally, using n threads for a stage reduces the average delay of the stage by a factor of n.

If threads in one stage *wait* for threads in the next stage to complete, the overall benefits of concurrency are greatly reduced. If the Sync Adapter thread waits for a response from the Sync Engine thread, which in turn waits for a response from the datastore thread, the system overall slows down appreciably. Communication between threads should be asynchronous. A thread in a precursor stage should be *called back* when results from a thread in a successor stage are available. Languages such as Java natively support threads and asynchronous communication. Similar thread libraries are available for languages such as C and C++. There may be situations when different stages of the pipeline are implemented in different languages or executed in different processes, or in different machines altogether. In such situations, it may be beneficial to use message-queue-based communication, such as IBM MQSeries® or Microsoft Message Queue.

Load balancing among different physical Servers is another way of achieving concurrency and high throughput. Load-balancing SyncML Servers is discussed in Chapter 4. The reader should note that different stages of the SyncML Server could also be load-balanced, with multiple machines dedicated to support a particular stage.

Failure and Recovery

Several failures can occur in an end-to-end data synchronization system. The Clients can fail, communication can fail, back-ends can fail, and the Server itself may crash or fail. In SyncML, the Server is expected to facilitate recovery from Client and communication failures. The Server maintains synchronization session information for each Client. The Server can detect Client failures by detecting discrepancies between sync anchors and initiate a slow synchronization with a Client. The Server can also use the stored session information to restart an ongoing synchronization session with a Client after a communication failure. The Server should be designed to mask transient back-end failures by retrying back-end operations or temporarily buffering back-end operations.

It is also important that the Server be prepared for the eventuality that it can crash or fail. State information such as capabilities, session, and identifier mapping should be in persistent storage or be backed up in persistent storage regularly. Using multiple machines per stage not only distributes load effectively but also can help continued operations when certain machines fail. For certain SyncML operations (such as Atomic, which has transactional semantics), the Server may be required to implement checkpoint and recovery functions.

13

Interoperability Verification

For a seamless end-user experience and for a successful standard, it is important that Clients and Servers be interoperable. To ensure this, SyncML® has introduced interoperability testing. This test must be passed in order to use the label SyncML conformant and the SyncML Interoperability Logo. The test ensures that products aspiring to be SyncML conformant and Interoperable actually conform to the specifications and interoperate with other implementations. This chapter covers the steps required to pass the SyncML Conformance [SCTP02] and Interoperability Testing Process [SITP02].

The goal of SyncML is to provide global data synchronization. It could be jeopardized if a consumer with a SyncML conformant Client is unable to synchronize with a SyncML conformant Server (assuming they are using the same communication bearer and the same data types).

It is the goal of SyncML to ensure the interoperability of conformant devices. To ensure this interoperability, SyncML has established the SyncML Interoperability Committee (SIC). The SIC focuses only on ensuring this interoperability.

SyncML has also established the following processes and tools to support this goal:

- Conformance Testing Process
- SyncML Conformance Test Suite
- Interoperability Testing Process with SyncFest™
- SyncML Interoperability Reference Pool

Getting the SyncML Logo is a multistep process (see Figure 13–1). The process applies for data synchronization (SyncML DS), as well as

244 Chapter 13 ▸ Interoperability Verification

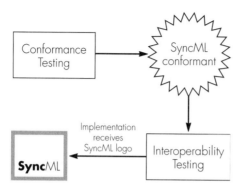

Figure 13–1
Interoperability Verification process

for Device Management (SyncML DM). In case of a device that supports both SyncML DS and SyncML DM, the interoperability of each protocol is verified separately. Interoperability testing for SyncML DS and SyncML DM can be done at the same SyncFest, so there is not much extra work involved.

Conformance Testing

The first step to obtain a SyncML Logo is to successfully complete the Conformance Testing process. It ensures that the Client or Server conforms to the appropriate SyncML DTD. For Device Management applications, the conformance process ensures that the Client or Server conforms to SyncML DM. In addition it ensures conformance to the supporting DTDs, such as the Device Information DTD [SDI02] and the Meta Information DTD [SMI02], as well as the supported transport binding specifications. The Conformance Testing also covers the Protocol–either the SyncML Synchronization Protocol [SSP02] or the SyncML Device Management Protocol [SDM02].

Each manufacturer can apply to use the term "SyncML Conformant" by going through a self-certification exercise and submitting the documents to the SIC. The SIC reviews the documents and lets the submitter know if his implementation is SyncML Conformant. The manufacturer is then free to use the term "SyncML Conformant" in relation to that one product and should continue with the Interoperability Testing process. The only valid reason not to continue is that there are no other devices to test against. In this case it is not possible to pass Interoperability Testing, because there are no other devices to test with.

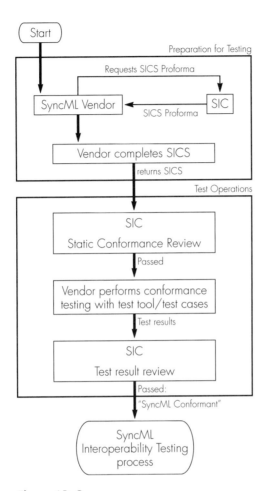

Figure 13-2
Conformance Testing process

SIC publishes the submitted documents on the SyncML Web site, but only after the device passes Conformance Testing.

This self-certification process consists of:

- Completing and submitting the SyncML Implementation Conformance Statement Proforma [SICSP02]
- Testing the implementation with the SyncML Conformance Test Suite
- Executing the manual test cases as described in the SyncML Manual Test Cases Document [SMTC02]
- Submitting all test results to SIC for review

An implementation that has successfully passed SyncML Conformance Testing, as shown in Figure 13-2, can use the term "SyncML conformant", but not the SyncML Logo. The SyncML Logo is reserved for products that successfully pass the SyncML Interoperability Testing process.

SyncML Implementation Conformance Statement (SICS)

A SICS proforma is available for download on the SyncML Web site. The proforma is a template that can be used by a manufacturer to check if an implementation is conformant. The completed SICS needs to be submitted to the SyncML Interoperability Committee. First, the manufacturer has to fill out the device information: Is the implementation a SyncML Server or Client, which are the supported content formats, and which are the transport bindings used (HTTP, WSP, or OBEX)? Contact information is also needed. Depending on whether the product is a Client or Server, the appropriate sections in the SICS need to be completed.

The SICS lists all the commands of the SyncML DTDs (both DS and DM), the supporting DTDs, the possible protocol element types, and the allowed MIME headers. Each feature is indicated as mandatory to support (MUST) or optional (MAY, SHOULD).

To be SyncML conformant, all MUST features in the appropriate sections have to be implemented. Even if only one MUST feature is not implemented, the SIC has to return the application to the submitter and notify them that their implementation is not conformant. From a conformance process point of view, the manufacturer is free to support or not support a feature that is marked with SHOULD or MAY. The SyncML Initiative strongly encourages manufacturers to implement the SHOULD features, whereas MAY features really are optional.

SyncML Conformance Test Suite (SCTS)

Based on the fact that interoperability is key for the success of SyncML in the market and with customers, the Initiative decided in the Fall of 2000 to develop a SyncML Conformance Test Suite (SCTS). It serves two purposes: First, it helps the developers of SyncML software to verify their implementations; second the SCTS helps during the SyncML Conformance Testing to easily verify if an implementation is conformant or not.

In November 2000, a team of software engineers started to work on the SCTS for the SyncML DTDs and protocols. First, they created a list of test cases that the Test Suite should execute, which was extensively reviewed by the SyncML Initiative before the actual implementation started. From February to September 2001, the tool went through extensive testing by all the companies that applied for SyncML conformance. Starting in November 2001, it was a requirement to pass the SCTS to get the SyncML Conformance label.

In September 2001, the Device Management Expert Group started to work on the test cases for a version of SCTS that tests conformance to the SyncML DM DTD and Protocol.

The development expenses for the SCTS are shared between the companies using it. A nonmember company has to pay the full license, Supporter-level members get a discount (equivalent to their membership fee), Promoter-level members get an even higher discount, and for Sponsor-level members, the license fee for SCTS is included in their membership fee. Currently, a SCTS license is bound to the major version of SyncML DS or SyncML DM that the purchased SCTS version tests. This license entitles company-wide usage, as well as updates. The up-to-date terms and conditions for SCTS are available at the SyncML Web site.

The SyncML Conformance Test Suite contains the following:
- SyncML Conformance Test Tool
- Test Cases
- Online Help
- List of changes

How to use the SCTS 2.x

SCTS version 2.x is a Windows® application. SCTS can act as a SyncML Server or as a SyncML Client. The SCTS contains the following six pages:
- General
- Objects
- Test
- Commands
- Logs
- Status

The first step is to configure SCTS to work with your Client or Server. SCTS comes with two sample configurations, one as a Client and one as a Server. These samples can also help one to configure the tool.

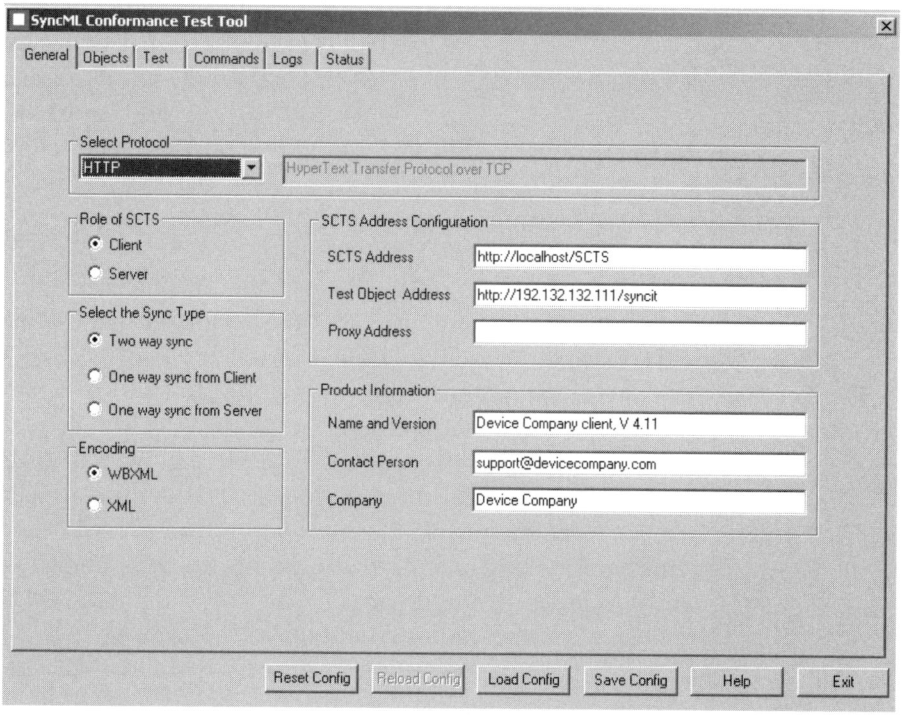

Figure 13-3
SyncML Conformance Test Tool—General page

The best way is to start is with the General page (Figure 13-3) and select whether SCTS should act as a Client or a Server. Usually the default settings (HTTP, two-way sync, WBXML) can be used.

SCTS supports communication over HTTP, as well as over OBEX. You will need to set the Address Configuration group. If SCTS is to act as a Client using HTTP, you will need to set the SCTS Address to the address of the local machine (e.g. localhost) plus some identifying string (in the example, SCTS). The Test Object Address is the full URI of the device to be tested. If a proxy is being used, then one must fill in the full URI of the proxy as well. If SCTS is to act as a Server using HTTP, only the SCTS Address needs to be filled in.

The entry fields in the product information group are added to the report that SCTS creates and are used by the SIC.

In case the SCTS is a Client, the next step is to enter the Server username and password on the Objects page (Figure 13-4) in the Source Details group. The username and password are used to access the SyncML Server.

Conformance Testing 249

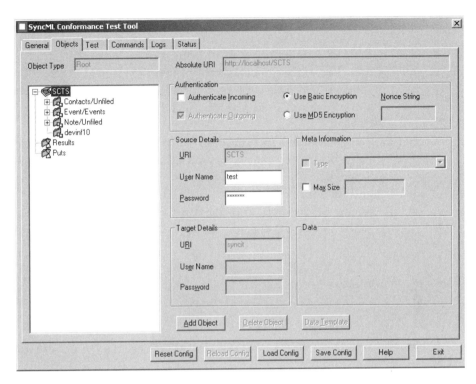

Figure 13–4
SyncML Conformance Test Tool—Objects page

Using the Auto Configure function on the Test page (Figure 13–5), SCTS can exchange the Device Information with the Server and use the data received to configure itself. The Auto Configure function is an easy way to configure SCTS, but it is still possible to enter all the require data manually on the Objects page.

The Objects page now lists the different content types that the Server supports, in this case, contacts, events, and notes. Note that an object called "devinf10" is shown, which represents the device information document that was exchanged. By clicking on one of these data types, the SCTS displays the respective Source and Target URIs, and the MIME type used to exchange data (for example text/x-vcard or text/calendar). After executing the different test cases, the SCTS also displays the stored source and target anchor on this page.

To pass the SCTS, it is necessary to go through one Test Group at a time in the indicated Test Group order. Each Test Group requires the previous Test Group to be executed–it is not possible to just execute

Test Group 6, for example. This is because the SCTS test cases need to set up some data on the Client or Server. SCTS also needs to be in a defined state before a Test Group can be executed.

For more general testing purposes, it is also possible to execute a random test case and start it without executing the previous ones. In some cases, this requires the SCTS to initiate a slow sync with the other device.

As shown in Figure 13–5, the SCTS marks the Test Groups that have been successfully passed and stops at the Test Group that failed. To each Test Group, SCTS describes the state that the device under test and the tool need to be in. For example, in Test Group #5 the datastore needs to be empty and for Test Group #6 five items need to be added to the device's datastore. Table 13–1 lists the various functions tested.

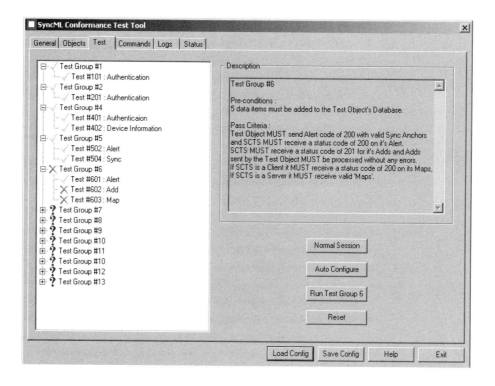

Figure 13–5
SyncML Conformance Test Tool—Test page

Table 13-1
Functions tested by SCTS

Test Group	Tested behavior
#1	Authentication with no credentials.
#2	Authentication with wrong credentials.
#3	Authentication with both Basic and MD5.
#4	Authentication with correct credentials and exchange of device information.
#5	Alert with slow sync and normal sync.
#6	Alert with Add of 5 objects and 5 corresponding Maps.
#7	Replace, Delete, Replace without target.
#8	Slow sync to check consistency of databases.
#9	Init and Sync in the same package, as well as maps to be sent in the next session.
#10	Maps sent from previous session. Add, Delete, Replace, and Status with multiple items.
#11	Slow sync to check consistency of databases.
#12	Slow sync with data only coming from the SCTS.
#13	Support of multiple messages.

The Command page (Figure 13-6) lists each step the SCTS has performed. The page lists, for example, the commands that were created. In addition it displays what was sent to and what was received from the test object. Each exchanged command is logged (including Session ID, Cmd ID, Msg ID, Command, Status, Source and Target Ref), as well as the result of the command.

The Logs page gives easy access to all log files generated by SCTS during test case execution.

Finally, every action and state is not only in the log files, but is also displayed on the status page (which is not shown here).

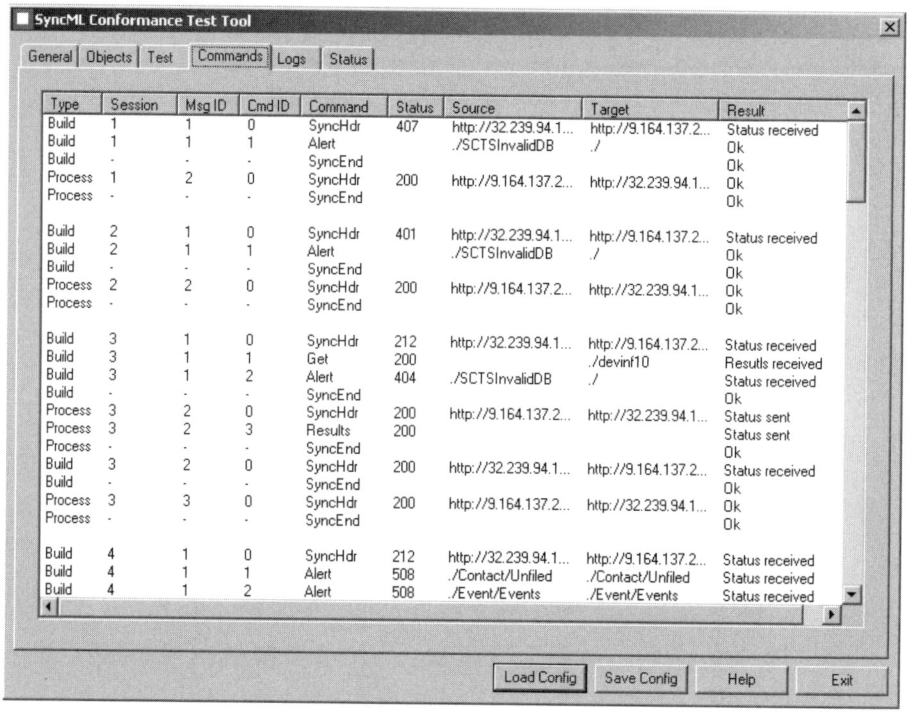

Figure 13-6
SyncML Conformance Test Tool—Command page

Interoperability Testing at SyncFest

The SyncML Initiative organizes five SyncFests a year. SyncFests take place in Europe, Asia, and North America. The SyncML Initiative provides a test location, and a preferred hotel. Internet access at the SyncFest (with externally accessible IP addresses) is also provided.

The SyncFest provides manufacturers with a convenient way to test the interoperability of their implementations with implementations from other manufacturers. A SyncML Client must successfully synchronize with two SyncML Servers from two other manufacturers before it is awarded the SyncML Logo. The same applies to Servers–they need to be tested with two Clients from other manufacturers. The SyncML Logo is licensed to manufacturers of interoperable products under the terms and conditions of the SyncML Logo Guidelines.

To be interoperable, a Client and a Server need to support the same content format, such as vCard 2.1, for example. As part of the

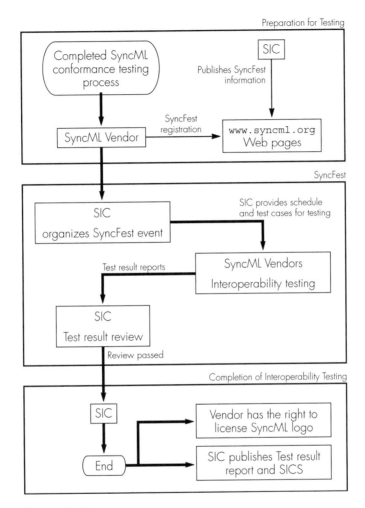

Figure 13-7
Interoperability Testing process

SyncML Representation Protocol [SRP02], SyncML defines content formats that Servers must support if the Server synchronizes certain data types. For example, for contacts a Server must support vCard 2.1 and (optionally) vCard 3.0.

Of course, there will be cases where somebody wants to use SyncML to synchronize some data types for which SyncML hasn't specified any content formats. For example, consider a Client that wants to synchronize multimedia data. The ideal approach would be to submit a change request to the SyncML Initiative and to suggest that a content format for this multimedia type be added to the SyncML Representation Specification. Even without this, as long as there are at least two Clients and two Servers from at least three different companies, the implementation can test at a SyncFest.

During a SyncFest, the SyncML Interoperability Committee schedules interoperability test sessions between Clients and Servers. The first priority is for devices that are not yet interoperable to test with devices that are already interoperable. This aids in debugging new implementations. The next priority is to test new devices from one manufacturer with new devices from another, and lastly to test compliant devices with compliant devices. If the SyncFest schedule permits, the goal is to also enable companies to test with as many implementations as possible.

The complete Interoperability Process is shown in Figure 13–7.

Virtual SyncFest

To provide a more flexible way for manufacturers to participate in Sync-Fests, the SyncML Initiative introduced Virtual SyncFests during the October 2001 SyncFest in San Francisco. This also gives SyncFest participants a wider selection of implementations to test with. This is limited to manufacturers with already interoperable devices.

This way, the Virtual SyncFest participants do not need to travel to a SyncFest. More important, devices can participate at a SyncFest even in cases were the SyncFest location does not provide the required wireless network infrastructure. For example, there is no 900 MHz GSM network, required by many European mobile phones, in North America.

SyncML Interoperability Reference Pool

To provide companies with the ability to test with SyncML compliant Clients and Servers between SyncFests, SyncML has established the SyncML Interoperability Reference Pool (SIRP). The SIRP contains several SyncML compliant Clients and Servers. These devices are from companies that are committed to provide others with a way to test against SyncML compliant products.

The goal of the SIRP is to enable companies to be better prepared for testing during a SyncFest. It is not a replacement for a SyncFest, though it is possible to get the Interoperability Logo just by testing with the SIRP.

Manufacturers who already have devices in the SyncML Interoperability Reference Pool can use the SIRP to do interoperability testing. To do so, the vendor submits a request to the SIC to select a number of devices from the SIRP that the applicant needs to test with to show interoperability. The minimum number of devices needed for testing is five, and all must be from different companies (excluding the applicant's company).

To receive the logo, the vendor must run the complete selection of test cases defined by SIC against at least three of the devices selected by the SIC.

To gain access to these devices, most Client devices, such as mobile phones, can just be purchased. Server manufacturers usually provide the ability to synchronize with their Servers over the Internet and provide special accounts to companies conducting tests.

Recertification

One of the most frequently asked questions is when an implementation needs to be recertified. Recertification is required as soon as the implementation goes through a major change–for example, it was updated to the next major release.

No recertification is required if the implementation only has minor changes or is executed on another operating system. Let's take a synchronization Server written in Java as an example. If this implementation was certified on one operating system and only minor adjustments had to be made to get the implementation working on the other operating system, then the SyncML Logo could also be used for the other operating systems. The same is true for a SyncML Client for an operating system that is included in several devices from the same or from different manufacturers (Symbian OS®, for example).

Part IV

SUMMARY AND THE FUTURE

14

Summary and the Future

SyncML® was initially created to provide a single open specification for data synchronization. Device management came as a second step for the SyncML Initiative. In both cases, companies and individuals realized the need for a single open standard, came together to produce the standard, and worked together to make these new standards as useful and interoperable as possible.

Ironically, like most standards, this standard should become invisible to the end-user. After all of the hard work to create a working standard, the end-user will probably never see the standards in action. They will become similar to the phone lines or water lines–they move the data from place to place and they "just work."

SyncML History

In order to properly understand the significance of the SyncML Initiative, perhaps a brief history is in order. The SyncML Initiative was publicly announced in February 2000. The purpose was to allow for the creation of an open standard for data synchronization. SyncML Data Synchronization 1.0 was published in December 2000. This was seen as a very good first effort, with a few important features left for later.

Another important area not yet addressed was device management. A group was formed within SyncML in May 2001. The group was formed from companies that were not only Sponsors of SyncML, but also of a new rank, Promoters. Promoters were companies that were not on the Board of SyncML, but wanted to work on the specifications.

Previously, only Sponsor companies were allowed to work on the specifications.

In February 2002, version 1.1 was released, including updates to data synchronization, and introduced device management. The data synchronization release included support for large object delivery and a split of the Representation Protocol Specification into two documents: a Common Representation document and a Data Synchronization Usage Representation document. The device management release also used the Common Representation document as a basis for their Device Management Usage Representation document.

The SyncML Initiative was not satisfied with just publishing a specification, though. In addition to the specifications, an interoperability process was defined and implemented. The first test of the interoperability process took place in February 2000, at the ETSI facility in Sophia Antipolis, France. The first SyncFest took place in April 2001 in Dallas, Texas. SyncML also created a semiautomated test suite to make conformance testing easier and more reliable. This test suite, called the SyncML Conformance Test Suite, or SCTS, was built with the intention of testing both Clients and Servers. SCTS development started in November 2000 and was later made a requirement for earning the SyncML Compliance mark.

Current Market Status

The number of SyncML compliant products is steadily increasing. A quick glance at the SyncML Web page (*www.syncml.org*) shows a healthy collection of Clients and Servers. Many of these Clients and Servers continue to test at SyncFests, both for confirmation of compliance and to assist newer products.

More SyncML Data Synchronization (DS) products are coming into the marketplace as SyncML becomes viewed as necessary technology. Both 3GPP™ and WAP Forum® consider SyncML DS to be important technology and have adopted its use.

SyncML Device Management (DM) will only accelerate this process. DM is viewed as necessary technology by both mobile carriers and large corporate IT departments.

Future SyncML Activities

SyncML is working towards the adoption of SyncML DM by the major standards organizations, such as 3GPP, WAP Forum, and OSGi™. Other standards organizations, such as the Java Community Process™ (JCP), are also being investigated. The purpose of this adoption is to make synchronization and device management a simple process—by having a single, open standard for these activities.

There are other possibilities for SyncML beyond data synchronization and device management. It is possible that the base technology could be applied to other areas, such as:

- Digital Rights Management—allowing users to buy, sell, and trade copyrighted materials. Organizations such as Napster™ indicate the need to implement a usable Digital Rights mechanism.
- SyncML Replication—allowing servers to share the data needed for client/server synchronization.
- Relational Data—allowing clients to have subsets of Relational tables.

The SyncML Initiative is also working on improving the existing specifications. Newer versions are planned with a typical delivery rate of one release per year. SyncML DS will be working on (a nonexclusive list):

- Improving the Device Information structure
- Adding additional authentication types, including Certificates and Challenge-Response
- Improved Transport Bindings to include more details on secure transports, and message chunking in WSP and OBEX
- Improvements on initial sync, including ease of use for first time synchronization

SyncML DM is working on (a nonexclusive list):

- Improvements on the Device Description Framework
- Improvements on initial Server contact with a Client
- Making changes to the SCTS to test SyncML DM Clients and Servers
- Software Delivery and Management

SyncML has also discussed other improvements, but has generally decided that most of these are too significant to add into a 1.x release and will be reconsidered for a later major release. In particular, SyncML has discussed the use of XML Schemas, and decided that it would produce too much of a change in the parsers and underlying technology. Likewise, other binary encoding schemes, such as XHTTP, will be considered for the next major release.

Of course, if problems arise, Erratas will also be published. Test cases (both manual and SCTS) are continually being refined. The specifications are always being reviewed for errors and possible clarifications. Naturally, if someone finds an error or an enhancement, they should send email to admins@syncml.org.

Future Markets

Products containing both SyncML DS and DM will start to appear as regular features in mobile devices, due to the influence of the various standards organizations requirements.

- Mobile phones will use SyncML DS to maintain the address book. Users will no longer have to worry about losing their data when they lose their phone or how to get their information onto a new phone.
- Mobile phones will use SyncML DM to maintain the user preferences and network connections. Carriers will be able to maintain a large number of mobile devices with much less difficulty. Users will be able to get better support when their devices have problems.
- Small, portable computers will use SyncML DS to maintain a large suite of data, ranging from address books and calendars to files on the local disk. Users will be able to access their data from anywhere and not worry about keeping everything up to date.
- Web Servers have already started to incorporate SyncML DS in their offerings. Carriers will start to offer SyncML DS and DM as part of their services.
- Larger computers will use SyncML DS to distribute user information to portable devices, in addition to the Web interface for the data. The user's data will be safely and securely synchronized in a seamless fashion.

SyncML will appear in many different markets, some faster than others. Slower acceptance in some markets may be due to a thorough review by a standards body or the proprietary nature of the market.

Of course, it is always difficult to guess the future. No doubt, new product areas will discover new uses for SyncML technology.

Part V
APPENDICES

Appendix A
Bibliography

This bibliography provides a listing of some potentially useful reference works. They are separated into the following categories: books, standards, RFCs and other useful websites.

Books

ACS+00: *Professional WAP.* Charles Arehart et al. Wrox Press, Inc. 2000.

Bol01: *Pure CORBA.* Fintan Bolton. Sams, 2001.

GS96: *Distributed Computing with IBM MQSeries.* Leonard Gilman, Richard Schreiber. John Wiley & Sons, 1996.

Gro01: *Java RMI.* William Grosso. O'Reilly, 2001.

HM01: *XML In a Nutshell: A Desktop Quick Reference.* Elliotte Rusty Harold, W. Scott Means. O'Reilly, 2001.

MB01: *Bluetooth Revealed: The Insider's Guide to an Open Specification for Global Wireless Communications.* Brent A. Miller, Chatschik Bisdikian. Prentice Hall, 2001.

Ree00: *Database Programming with JDBC and Java.* George Reese. O'Reilly, 2000.

Roc98: *DCOM Explained.* Rosemary Rock-Evans. Digital Press, 1998.

San98: *Hands On ODBC Developer's Guide.* Roger Sanders. McGraw Hill, 1998.

SyncML Specifications

Note that the ordering is the suggested reading order for these documents.

Data Synchronization Documents

SSP02: SyncML Synchronization Protocol.
http://www.syncml.org/download/syncml_sync_protocol_v11_20020215.pdf

SRP02: SyncML Representation Protocol.
http://www.syncml.org/download/syncml_represent_v11_20020215.pdf

SDS02: SyncML Representation Protocol, Data Synchronization Usage.
http://www.syncml.org/download/syncml_sync_represent_v11_20020215.pdf

SDI02: SyncML Device Information DTD.
http://www.syncml.org/download/syncml_devinf_v11_20020215.pdf

SMI02: SyncML Meta Information DTD.
http://www.syncml.org/download/syncml_metinf_v11_20020215.pdf

SHB02: SyncML HTTP Binding.
http://www.syncml.org/download/syncml_http_v11_20020215.pdf

SOB02: SyncML OBEX Binding.
http://www.syncml.org/download/syncml_obex_v11_20020215.pdf

SWB02: SyncML WSP Binding.
http://www.syncml.org/download/syncml_wsp_v11_20020215.pdf

Device Management Documents

SDP02: SyncML Device Management Protocol.
http://www.syncml.org/download/syncml_dm_protocol_v11_20020215.pdf

SDM02: SyncML Representation Protocol, Device Management Usage.
http://www.syncml.org/download/syncml_dm_represent_v11_20020215.pdf

SBO02: SyncML Device Management Bootstrap.
http://www.syncml.org/download/syncml_dm_boot_v11_20020215.pdf

SCO02: SyncML Device Management Conformance Requirements.
http://www.syncml.org/download/syncml_dm_conreqs_v11_20020215.pdf

SNS02: SyncML Device Notification Initiated Session.
http://www.syncml.org/download/syncml_dm_notification_v11_20020215.pdf

SSE02: SyncML Device Management Security.
http://www.syncml.org/download/syncml_dm_security_v11_20020215.pdf

SSO02: SyncML Device Management Standardized Objects.
http://www.syncml.org/download/syncml_dm_std_obj_v11_20020215.pdf

STD02: SyncML Device Management Tree and Description.
http://www.syncml.org/download/syncml_dm_tnd_v11_20020215.pdf

Interoperability Documents

SCTP02: SyncML Conformance Testing Process.
  http://www.syncml.org/interoperability/testing_process.pdf

SICSP02: SyncML Interoperability Conformance Statement Proforma.
  http://www.syncml.org/interoperability/proforma.doc

SITP02: SyncML Interoperability Testing Process.
  http://www.syncml.org/interoperability/interop_process.pdf

SMTC02: SyncML Manual Test Cases.
  http://www.syncml.org/interoperability/test_cases.pdf

Standards

BlIR01: Bluetooth SIG, IrDA Interoperability–Bluetooth Core Specification.
  http://www.bluetooth.com

DOM02: Document Object Model. World Wide Web Consortium.
  http://www.w3c.org/DOM

IrMC00: IrDA Infrared Mobile Communications.
  http://www.irda.org/standards/specifications.asp

OBEX99: IrDA Object Exchange Protocol.
  http://www.irda.org/standards/specifications.asp

SAX02: Simple API for XML. Saxproject.
  http://www.saxproject.org

UWMC01: USB Implementers Forum, USB CDC Subclass Specification for Wireless Mobile Communications Devices.
  http://www.usb.org

VCAL: IMC, vCalendar–Electronic Calendaring and Scheduling Exchange Format.
  http://www.imc.org/pdi/vcal-10.doc

VCARD21: IMC, vCard–The Electronic Business Card.
  http://www.imc.org/pdi/vcard-21.doc

WBXML01: WAP Binary XML Content Format Specification.[1]
  http://www.wapforum.org/what/technical.htm

WLAN02: IEEE 802.11 Wireless Local Area Networks.
  http://grouper.ieee.org/groups/802/11/

---

1. The WAP Forum has been subsumed by the Open Mobile Alliance, Ltd. All WAP documents may be found at their Web site: http://www.openmobilealliance.org.

WPR01: WAP Forum, WAP Provisioning Architecture Overview.
http://www.wapforum.org/what/technical.htm

WPU01: WAP Forum, WAP Push Architecture Overview.
http://www.wapforum.org/what/technical.htm

WSP01: WAP Forum, Wireless Session Protocol Specification.
http://www.wapforum.org/what/technical.htm

WTLS01: WAP Forum, Wireless Transport Layer Security.
http://www.wapforum.org/what/technical.htm

RFCs

RFC0822: Standard for the format of ARPA Internet text messages. (Obsoleted by RFC2822)
http://www.ietf.org/rfc/rfc0822.txt

RFC1157: Simple Network Management Protocol.
http://www.ietf.org/rfc/rfc1157.txt

RFC1213: Management Information Base for network management of TCP/IP-based internets.
http://www.ietf.org/rfc/rfc1213.txt

RFC1321: MD5 Message-Digest Algorithm.
http://www.ietf.org/rfc/rfc1321.txt

RFC1945: Hypertext Transfer Protocol–HTTP/1.0.
http://www.ietf.org/rfc/rfc1945.txt

RFC2045: Multipurpose Internet Mail Extensions (MIME) Part One: Format of Internet Message Bodies.
http://www.ietf.org/rfc/rfc2045.txt

RFC2046: Multipurpose Internet Mail Extensions (MIME) Part Two: Media Types.
http://www.ietf.org/rfc/rfc2046.txt

RFC2068: Hypertext Transfer Protocol–HTTP/1.1. (Obsoleted by RFC2616)
http://www.ietf.org/rfc/rfc2068.txt

RFC2069: An Extension to HTTP: Digest Access Authentication. (Obsoleted by RFC2617)
http://www.ietf.org/rfc/rfc2069.txt

RFC2104: HMAC: Keyed-Hashing for Message Authentication.
http://www.ietf.org/rfc/rfc2104.txt

RFC2119: Key words for use in RFCs to Indicate Requirement Levels.
http://www.ietf.org/rfc/rfc2119.txt

RFC2246: Transport Layer Security.
http://www.ietf.org/rfc/rfc2246.txt

RFC2426: vCard MIME Directory Profile.
http://www.ietf.org/rfc/rfc2426.txt

RFC2445: Internet Calendaring and Scheduling Core Object Specification (iCalendar).
http://www.ietf.org/rfc/rfc2445.txt

RFC2459: Internet X.509 Public Key Infrastructure Certificate and CRL profile.
http://www.ietf.org/rfc/rfc2459.txt

RFC2451: Lightweight Directory Access Protocol (v3).
http://www.ietf.org/rfc/rfc2451.txt

RFC2510: Internet X.509 Public Key Infrastructure Certificate Management Protocols.
http://www.ietf.org/rfc/rfc2510.txt

RFC2511: Internet X.509 Certificate Request Message Format.
http://www.ietf.org/rfc/rfc2511.txt

RFC2585: Internet X.509 Public Key Infrastructure Operational Protocols: FTP and HTTP.
http://www.ietf.org/rfc/rfc2585.txt

RFC2616: Hypertext Transfer Protocol–HTTP/1.1.
http://www.ietf.org/rfc/rfc2616.txt

RFC2617: An Extension to HTTP : Digest Access Authentication.
http://www.ietf.org/rfc/rfc2617.txt

RFC2817: Upgrading to TLS Within HTTP/1.1.
http://www.ietf.org/rfc/rfc2817.txt

RFC2818: HTTP Over TLS.
http://www.ietf.org/rfc/rfc2818.txt

RFC2821: Simple Mail Transfer Protocol.
http://www.ietf.org/rfc/rfc2821.txt

RFC2822: Internet Message Format.
http://www.ietf.org/rfc/rfc2822.txt

RFC1149: Standard for the transmission of IP datagrams on avian carriers.
http://www.ietf.org/rfc/rfc1149.txt

Other Useful Web Sites

The SyncML Initiative
http://www.syncml.org

WAP Forum
http://www.wapforum.org

3GPP–Next Generation Partnership Program
http://www.3gpp.org

IETF–Internet Engineering Task Force
http://www.ietf.org

IMC–Internet Mail Consortia
http://www.imc.org/pdi

IrDA–Infrared Data Association
http://www.irda.org

OMA–Open Mobile Alliance
http://www.openmobilealliance.org

Appendix B
Glossary

This section provides a glossary of most of the terms used in the book.

Glossary

Access Control List A table that indicates the access rights a user has for a particular system object.

Asynchronous Request-Response Protocol A message-exchange protocol between two computers where one party issues a request and another party optionally responds to the request, with the requesting party not waiting for the response. If the requesting party expects a response, it either checks for a response repeatedly or is alerted when a response is ready. A typical example of this protocol is SMTP.

Authentication The process of determining whether someone or something is, in fact, who or what it is declared to be.

Database Transaction Processing A sequence of information exchange and related work that is treated as a unit for the purposes of satisfying a request and for ensuring database integrity.

Bootstrapping The process by which the unconfigured mobile device is taken from the initial state to the minimally configured state.

Certificates A digital certificate is an electronic object that establishes credentials. It is issued by a certification authority. It contains a name, a serial number, expiration dates, a copy of the certificate holder's public key (used for encrypting messages and digital signatures), and the digital signature of the certificate-issuing authority so that a recipient can verify the certificate.

Change Log A data structure that stores the history of operations made to data items in a datastore. A timestamp may be stored along with each operation.

Client Implementation An implementation of SyncML that operates in the Client Role.

Client Software The software that implements the SyncML Client Role.

Content Format The format of the content, e.g. vCard or vCalendar.

Data Synchronization The process by which copies of data shared between two or more computers is kept consistent.

Datastore The storage location of data, e.g. a Database or a file.

Database Transaction Processing A sequence of information exchange and related work that is treated as a unit for the purpose of satisfying a request and for ensuring database integrity.

Device Description Framework SyncML DDF DTD used for describing data within the Management Tree.

DM Client A Client that implements the SyncML Device Management Protocol.

DM Server A Server that implements the SyncML Device Management Protocol.

Field Mapping The process of correlating fields (or columns) in a data record in the datastore of one computer with those in another computer.

Footprint (Static) The size of the object code or that of the interpreted bytecode for a program or a system.

Footprint (Dynamic) The overall memory required when code is executed. This is dependent on an actual execution instance.

Identifier (ID) Mapping The process of correlating identifiers of data items in the datastore of one computer with those in another computer.

Integrity Violation A condition where an operation on a data item contradicts certain constraints on the value of the data item. A typical example is updating some data value out of a prescribed range.

Load Balancing The process by which workload is equally (or close to equally) distributed among a set of computers.

Local Synchronization Data synchronization over a short-range (proximity-based) network connection such as a serial cable, infrared, or Bluetooth™. This type of synchronization typically occurs between a mobile device and a personal computer.

Management Tree The mechanism used by SyncML DM to access individual items on a DM Client.

Management Object The entities that can be manipulated by SyncML DM Servers.

Many-to-one Synchronization A synchronization topology that consists of a group of two or more entities, where there is only a single entity with which the other entities synchronize data. A typical example is a PDA and a laptop computer synchronizing with a desktop computer. This topology is also called the Star topology or the Central Master topology. Note that one-to-one synchronization is a special case of this topology.

Many-to-many Synchronization A synchronization topology that consists of a group of two or more entities, where any entity can synchronize data with any other entity. A typical example is a PDA, a laptop computer, and a desktop computer synchronizing with each other. This topology is also called the Peer-to-Peer topology. Note that Many-to-one and One-to-one topologies are special cases of this topology.

Multiple-Path Synchronization The context of many-to-one synchronization where all accesses to the central datastore do not pass through the synchronization Server.

One-to-one Synchronization A synchronization topology that consists strictly of a two-member group, where the members only synchronize data with each other. A typical example is a PDA that only synchronizes with a specific the personal computer and vice-versa. This topology is also called the dedicated-pair topology.

One-way Synchronization Data synchronization between two entities where only one entity informs the other about data updates.

Remote Synchronization Data synchronization over a longer-range (infrastructure-based) network connection such as a wireline/wireless local/wide area network. This type of synchronization typically occurs between a mobile device or a personal computer and a server.

Server Alert The process a Server sending an Alert or Trigger to a client.

Service Provider An entity that provides internet-enabled applications such as email, contacts, and calendar-management. Typically, the services are made available using a web server or a web portal.

Servlet A server-side Java™ application invoked by an HTTP server that handles requests from internet clients

Single-Path Synchronization The context of many-to-one synchronization where all accesses to the central datastore pass through the synchronization Server.

Slow Synchronization The process of synchronization where one entity exchanges the values of all items in a datastore with another entity. This is typically performed to recover from failures when it is not possible to determine the last common synchronization anchor between two entities.

Synchronization Analysis The process of analyzing changes made to data items by two synchronizing entities, detecting conflicts (changes made to the same data item by both entities), and possibly resolving conflicts.

Synchronous Request-Response Protocol A message-exchange protocol between two computers where one party issues a request and another party responds to the request, with the requesting party waiting until the response arrives. A typical example of this protocol is HTTP.

Synchronization Vendor A software vendor that specializes in developing software for synchronizing data between applications on different computers. Typically, synchronization vendors are not the original vendors of the applications.

Sync Anchor A marker associated with a datastore and one of its synchronizing partners, logically indicating the last instance of synchronization with that partner. The marker can be a timestamp or a logical counter.

Sync Type One of the supported synchronization scenarios. E.g. Two-way Sync, Slow Sync.

SyncML Client (Client) An implementation of the SyncML Data Synchronization Protocol, specifically implementing the Client Role.

SyncML Component One of the SyncML DTDs or Specifications.

SyncML Entity (Client/Server) A device that implements the SyncML specifications (in either a Client or Server Role or both).

SyncML Framework The Framework encompasses the Representation Protocol, the Synchronization Protocol, the Transport Binding and the SyncML Adapter. It does not include any Sync Agent(s), Sync Engine(s) or associated applications.

SyncML Message A Message is a a well-formed XML document that contains the SyncML Header and Body. SyncML Messages are used to communicate the necessary synchronization information.

SyncML Package A synchronization session consists of SyncML Packages being exchanged between a Client and a Server. A SyncML Package may be split into multiple SyncML Messages, if there are size constraints.

SyncML Server (Server) An implementation of the SyncML DS Protocol, specifically implementing the Server role.

Timestamp The actual time of an operation on a datastore on an individual data item or a group of data items. Many synchronization systems store the actual time of an operation along with the operation itself to select which updates are appropriate for a particular synchronization session.

Two-way Synchronization Data synchronization between two entities where the entities exchange information about data updates that occurred in each entity.

Transcoding The process of altering the format of data items to make them suitable for the characteristics of a mobile device.

Transport Binding The specification of how a higher-level protocol (e.g., a synchronization protocol) utilizes a specific transport protocol (e.g., HTTP). For example, in the context of SyncML® and HTTP, this entails specifying exactly how SyncML messages are encapsulated and transferred using the format and message-exchange sequence prescribed by HTTP.

Uniform Resource Indicator (URI) A uniform method of object location within the Internet. URLs (Universal Resouce Locator) and URNs (Universal Resouce Name) are special cases of URIs.

Version Vector An array of logical counters associated with a data item in a datastore, where the value of the ith counter indicates the version of the data item in the ith peer in a peer-to-peer synchronization topology.

Appendix C
Trademarks

This section lists trademarked terms used in the book.

3GPP™
Anyday.com™
Bluetooth™
excite@Home™
EPOC™
Ericsson®
FoneSync™
IBM®
Intellisync Anywhere®
IrDA®
Java™
Linux™
Lotus®
Lotus Domino™
Lotus Notes®
Matsushita®

Metrowerks®
Microsoft®
Microsoft Exchange®
Microsoft Outlook®
Motorola®
MQSeries®
Netscape®
Nokia®
Open Mobile Alliance®
Open Mobile Architecture®
Openwave®
Palm®
Palm OS ®
Palm Pilot™
PocketPC™
Psion®
Starfish®
Symbian®
Symbian OS®
SyncML®
TrueSync®
Universal Serial Bus™
Unwired Planet®
Visual Basic®
WAP™
WAP Forum®
Windows®
Wireless Village®
XTndConnect®
Yahoo!®

Index

Numerics

2G 57
3G 56, 57
3GPP 16, 17, 261

A

Access Control List 182
Add 29, 106, 112, 176, 212
Alert 86, 116, 178, 212
Application Type 215
Architecture 55, 220
Asynchronous Request-response 60
Atomic 29, 178, 241
Authentication 83, 90, 153, 180, 213

B

Base64 156
Basic Authentication 84, 154
Bluetooth 9, 25, 31, 56, 57, 59, 71, 74, 151, 165

Bookmark 216
Bootstrapping 170, 173, 174

C

C 214
C++ 214
Calendar 146, 207, 216, 217
Callback Functions 189, 190, 192, 227
Cataloging-in-Publication Data ii
Central Master Topology 5
Certificate 225, 237
Chal 121, 142
Change Bit 67
Change Log 58, 65, 66, 213
Client 79, 83, 85, 136, 141, 144, 205, 206, 207, 210, 211, 213, 215, 218, 219, 220
Client Engine 209, 210
Client Modification 89
Cluster Topology 7
Code Page 142
Code Space 139, 142
Command 29

Command Dispatcher 190
Completion Phase 82, 89
Conflict 37, 39, 46, 49, 89, 101, 223, 226, 227, 231, 232, 233, 235, 238, 239
Conflict Resolution 16, 209
Conflict Update 14
Conformance Testing 243–6
Conformity Check 83, 87
Connect Operation 151
Constraint Violation 13, 235
Consumer 37
Contacts 146, 207, 216, 217
Content Type 146
CORBA 33
CTCap 145
CTType 145
Customer Care 165

D

Data 131
Data Exchange 88
Data Exchange Phase 82, 83
Data Management 222, 223, 234
Data Synchronization 21, 23, 24, 25, 30, 33, 34, 259, 261
Datastore 145, 207, 218
Datastore Adapter 210, 218, 219, 228–9
DataType 145
DCOM 33
DDF 171, 172, 173, 181, 182, 183, 261
Decoder 189, 192
Dedicated Pair 5
defaultPrintFunc 194
Delete 29, 107, 114, 177, 212
Device Capabilities 83, 85
Device Information 145, 213, 249, 261
Device Information DTD 77, 78, 85, 135, 136, 138, 144, 145, 244
Device Management 163, 164, 166, 171, 173, 178, 181, 183, 259, 261
Device Management Protocol 47, 50, 138, 244
Device Manufacturer 206
Device Platform 205, 207, 213
DevID 145, 146
DevInf 145, 213, 249, 261

DevInf DTD 77, 78, 85, 135, 136, 138, 144, 145, 244
DevTyp 145, 146
Diagnostics 165, 167, 183
Digitial Rights Management 261
DisplayName 145
DM Bootstrap Protocol 173, 174
DM Client 173, 176, 177, 178, 179, 180
DM Protocol 170, 172, 173, 175, 178, 180, 183
DM Server 173, 176, 178, 179, 180, 181
DM Session 178
DMTF 166
DOM 63
DSMem 145
DTD 29, 77
Duplicate 96, 97
Dynamic Footprint 66

E

EDGE 18
Email 148, 216
EMI 140, 141, 142
Encoder 189, 192
Encryption 180, 181
Enterprise 35, 37, 48, 49, 50, 51, 222
Enterprise Server 68, 237
EPOC 192
Ericsson 17
Ethernet 56, 57
Ext 145

F

Field Mapping 226, 236
Filter 221, 227
Final 93, 123
FOMA 18
Footprint 210
Format 140, 141
Framework 135
FreeID 140, 141
FreeMem 140, 141
FwV 145

G

Get 85, 176, 212
Globally Unique Identifier 14
GNU Mingw 32 192
GPRS 18, 165
GSM 10, 142
GUID 14, 95, 101

H

Hierarchy Topology 8
HTML 216, 236
HTTP 27, 40, 42, 43, 45, 58, 59, 60, 61, 67, 71, 72, 74, 76, 136, 148, 149, 151, 155, 208, 210, 224, 238
HTTP Client 149
HTTP Server 149
HTTPS 157
HwV 145

I

iCalendar 62, 216
ID Mapping 89, 95, 96, 98, 102, 213
Identifier Mapping 228, 241
IEEE 802.11 56, 57
IETF 148
IMAP 60
Infrared 9, 21, 31, 150
Infrared Data Association (IrDA) 16, 59, 74, 150, 151
Infrared Mobile Communications (IrMC) 17, 25
Initialization Phase 82, 83, 85, 86, 90
Integrity 180, 181
Integrity Violations 70
Interoperability 77, 205, 216, 255
Interoperability Logo 243
Interoperability Testing 243
IrDA 16, 59, 74, 150, 151
IrMC 17, 25
Item 131

J

Java 213, 214
JCP 261
JDBC 228, 237
Journal 146

L

LAN 149
Large Object 91, 92, 147
Last 140, 141
Linux 192, 213, 214
Load Balancing 70, 241
Local Area Network 149
Local Synchronization 4, 21, 22, 25, 222
Locally Unique Identifier 15
LocName 123
LocURI 124
Lotus Domino 9
Lotus Notes 9
LUID 15, 95, 101

M

Maintenance Release 188
Man 145
Management Object 176, 177, 182
Management Tree 173, 176, 181, 182
Mandatory Features 212
Manual Testcases 245
Many-to-many Synchronization 4, 7, 56, 64, 65
Many-to-one Synchronization 4, 6, 57, 65
Map 102, 212
Mark 140, 141
Marker 86, 87
MaxGUIDSize 145, 147
MaxID 145
MaxMem 145
MaxMsgSize 140, 141
MaxObjSize 140, 141
MD5 32, 84, 155
MD5 Authentication 84
Mem 139, 140, 141
Memory Management 189

Message 28, 78, 83, 91, 92, 136, 137, 142, 148, 151
Message Digest 67, 71, 237
Message-based Authentication 84
Meta Information 140, 141, 142
Meta Information DTD 77, 78, 135, 136, 138, 139, 142, 146, 244
MetInf 140, 141, 142
MetInf DTD 77, 78, 135, 136, 138, 139, 142, 146, 244
Metrowerks 192
MIB 166, 171
Microbrowsing 149
Microsoft Exchange 9
Microsoft Outlook 9
MIDP 213, 214
MIME 61, 71, 75, 76, 140, 144, 146, 148, 149, 216, 226, 246
MMS 148
Mobile Computing 3
Mobile Software Platform 213, 214
Mod 145
Motorola 17
MP3 35
MQSeries 31
Multiple Messages 89, 91, 213, 225
Multiple-path Synchronization 229, 231, 232, 233, 234, 235, 237, 239

N

Name Space 139, 142
Next 140, 141
NextNonce 139, 140, 141, 142
Nokia 17
Nonce 84
NoResp 126
NumberOfChanges 127

O

OBEX 27, 59, 61, 71, 72, 74, 136, 148, 150, 151, 157, 210, 224
OBEX Client 152
OBEX Server 152

ODBC 237
OEM 145
One-to-one Synchronization 4, 5, 56
One-way Synchronization 30, 80, 81, 237
Open Database Connectivity (ODBC) 24
Openwave 17
Operator 206

P

Package 26, 28, 29, 31, 50, 82, 224, 225, 230, 238
Package #0 150
Palm 192
Palm OS 214
Parameter Configuration 165, 183
ParamName 145
Pass-thru Synchronization 4
Peer-to-peer 5
Performance 206, 210, 211
Pocket PC 192, 214
Policy 226, 227
POP3 60
Portal 22, 24
Post Method 149
Process Flow Command Elements 119
Programming Language 214
Promoter 25
PropName 145
Protocol Management 222, 223
Provisioning 169
Put 85, 212

Q

Quality 206

R

RAM 92, 210, 211
Read-Write Conflict 13
Recertification 255
Reference Implementation 204
Refresh 107

Relational Data 261
Remote Method Invocation (RMI) 31
Remote Synchronization 4, 10, 22, 23, 24
Replace 29, 106, 116, 177, 212
Representation Protocol 26, 27, 28, 29, 42, 71, 74, 77, 78, 103, 135, 136, 138, 142, 212, 223, 224
RespURI 127, 212
Results 212
Roles 30
Rx 145
Rx-Pref 145

S

SAX 64
SCR 133
SCTS 189, 243, 245, 251, 261
Search 29, 108, 117
Search Grammar 106
Secure Transport 158
Segmentation and Reassembly 92
Semantic Conflict 14
Sequence 178
Sequence Number 86
Serial Communication 21
Server 79, 80, 85, 101, 136, 141, 144, 150
Server Alert 150
Server Alert Phase 90
Server Alerted Sync 81, 82, 90, 149
Server Modification 89
Service Discovery 85
Service Provider 22, 24, 38, 39, 41, 42, 68, 206, 222
Servlet 224
Session-based Authentication 84
SessionID 128
SftDel 128
SharedMem 140, 141, 145
Short Message Service 150
SICS 243, 245, 246
SIM card 141
Single-path Synchronization 230, 231, 232, 233, 237
SIRP 243, 255
Size 140, 141, 145
Slow Synchronization 80, 81, 87, 94, 97, 98, 107, 147, 213, 237, 241
smlEndSync 196
smlInit 193
smlInitInstance 194
smlLockReadBuffer 197
smlProcessData 199
smlSetEncoding 194
smlStartAtomic 195
smlStartMessage 195
smlStartSync 195
smlString2Pcdata 197
smlTerminate 193
smlUnlockReadBuffer 198
SMS 150, 217
SMTP 31, 59
SNMP 166
Software Management 165, 183
Source 129
SourceRef 129, 145
Sponsor 25
SSL 67, 71, 158, 181
SSL Cipher Suites 159
SSL Handshake 159
Stale Read 13
Star Topology 5
State Machine 83
Static Conformance Requirements 133
Static Footprint 66
Status 29, 111, 212
Supporter 25
Supporter Summit 187
SupportLargeObjs 145, 147
SupportNumberOfChanges 145, 147
SwV 145
Symbian OS 192, 213, 214
Sync 118, 212
Sync Anchor 65, 70, 87, 139, 140, 213, 225
SyncBody 111
SyncCap 145, 147
Sync Engine 37
SyncFest 33, 252–3, 260
SyncHdr 110
Synchronization Agent 27, 33, 63
Synchronization Analysis 222
Synchronization Application 208
Synchronization Engine 27
Synchronization Protocol 26, 27, 29, 30, 42, 74, 77, 78, 80, 82, 83, 91, 93, 94, 97, 135,

138, 144, 150, 172, 209, 213, 223, 224, 225, 244
Synchronization Scenario 80
Synchronization Session 15, 82, 87
Synchronization Topologies 4, 55
Synchronization Vendors 22, 24, 27
Synchronous Request-response 60, 61
Sync Management 222, 223
SyncML Client 79
SyncML Conformance Test Suite 189, 243, 245, 251
SyncML Conformant 244
SyncML DTD 136, 142
SyncML Implementation Conformance Statement 245, 246
SyncML Initiative 150, 163, 166, 170, 183, 187, 259
SyncML Interoperability Committee 243, 245
SyncML Interoperability Reference Pool 243, 255
SyncML Server 79
Sync Type 80, 81, 83, 88, 90, 91, 144

Tx-Pref 145
Type 140, 141

U

UDP 150
UID 14
UMTS 10, 18
Unique Identifier 14
Universal Serial Bus 150
Unwired Planet 17
Update History 233, 235
URI 76, 86, 104
URN 76, 104
Usability 206, 210, 211
Usage Model 215
USB 9, 71, 150, 151, 165
USB Implementers Forum 151
User Interface 208
UTC 145, 146
UTRA 18

T

Target 130
Target Address Filter 105
TargetRef 130
TCP/IP 49, 58, 59, 60, 61, 72, 148, 149, 151
Third Generation Partnership Program 17
Timestamp 65, 67, 86
Timezone 146
TLS 158, 181
Todo 146, 207, 216, 217
Toolkit 50, 187, 204, 209, 225, 230
Toolkit Architecture 192
Toolkit Installation 192
Transcoding 236
Transport 189
Transport Binding 31, 244, 261
Triggers 235, 239
Troubleshooting 167, 170
Trusted Relationship 173, 175
Two-way Synchronization 30, 80, 81, 88, 99, 147, 213, 229, 237
Tx 145

V

ValEnum 145
vBookmark 216
vCalendar 31, 62, 73, 140, 216, 226, 228, 236
vCard 31, 42, 73, 216
VerCT 145
VerDTD 130, 145
VerProto 131
Version 140, 141
Version Vector 64, 65
vMessage 217
vNote 216

W

WAP 40, 41, 46, 57, 59, 71, 238
WAP Forum 17, 142, 149, 261
WAP Gateway 93, 150
WAP Provisioning 175
WAP Push 40, 149, 150, 173, 178
WBXML 57, 59, 76, 139, 142, 143, 155, 189

W-CDMA 18
Web Services 39, 40
Windows 192
Wireless Session Protocol (WSP) 59
WML 216
Workspace 191
World Wide Web Consortium 142
Write-Write Conflict 12
WSP 31, 42, 43, 45, 59, 60, 61, 68, 71, 74, 136, 148, 149, 210
WTLS 157, 181
WXBML 133

X

X.509 32
XHTML 216
XML 142
Xnam 145
xptBeginExchange 203
xptCloseCommuncation 203
xptEndExchange 203
xptGetProtocols 202
xptOpenCommunication 202
xptSelectProtocol 202
Xval 145

informIT

www.informit.com

YOUR GUIDE TO IT REFERENCE

Articles

Keep your edge with thousands of free articles, in-depth features, interviews, and IT reference recommendations — all written by experts you know and trust.

Online Books

Answers in an instant from **InformIT Online Book's** 600+ fully searchable on line books. Sign up now and get your first 14 days **free**.

POWERED BY

Catalog

Review online sample chapters, author biographies and customer rankings and choose exactly the right book from a selection of over 5,000 titles.

Wouldn't it be great

if the world's leading technical publishers joined forces to deliver their best tech books in a common digital reference platform?

They have. Introducing
InformIT Online Books
powered by Safari.

- **Specific answers to specific questions.**

InformIT Online Books' powerful search engine gives you relevance-ranked results in a matter of seconds.

- **Immediate results.**

With InformIt Online Books, you can select the book you want and view the chapter or section you need immediately.

- **Cut, paste and annotate.**

Paste code to save time and eliminate typographical errors. Make notes on the material you find useful and choose whether or not to share them with your work group.

- **Customized for your enterprise.**

Customize a library for you, your department or your entire organization. You only pay for what you need.

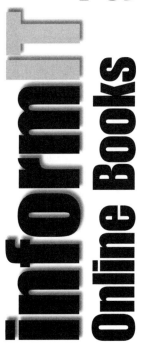

Get your first 14 days FREE!

InformIT Online Books is offering its members a 10 book subscription risk-free for 14 days. Visit **http://www.informit.com/onlinebooks** for details.

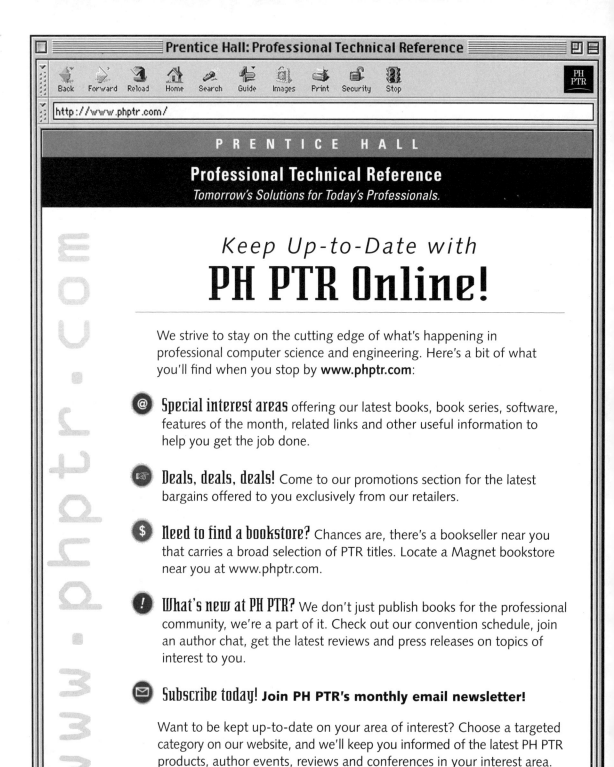